Programmable Logic
Handbook

Programmable Logic Handbook

Second Edition

Geoff Bostock

Newnes
An imprint of Butterworth-Heinemann Ltd
Linacre House, Jordan Hill, Oxford OX2 8DP

 A member of the Reed Elsevier plc group

OXFORD   LONDON   BOSTON
MUNICH   NEW DELHI   SINGAPORE   SYDNEY
TOKYO   TORONTO   WELLINGTON

First published by Collins Professional Books 1987
Second Edition published by Butterworth Heinemann Ltd 1993
Reprinted 1994

British Library Cataloguing in Publication Data
Bostock, Geoff
  Programmable Logic Handbook. – 2Rev. ed
  I. Title
  621.395

ISBN 0 7506 0808 0

Library of Congress Cataloguing in Publication Data
Bostock, Geoff.
  Programmable logic handbook/Geoff Bostock. – 2nd ed.
  p.   cm.
  Includes bibliographical references and index.
  ISBN 0 7506 0808 0
  1. Programmable logic devices.   I. Title.
  TK7872.L64B675   1993
  621.39′5–dc20   92–27737   CIP

Typeset by Vision Typesetting, Manchester
Printed in England by Clays Ltd, St Ives plc

Contents

Preface to the Second Edition

In the five years since the first edition was published there have been, as might be expected, many innovations in the subject of programmable logic. In particular there was very little reference in the first edition to LSI devices; these now warrant a chapter to themselves and, I would expect, this could easily be expanded to two chapters in a little while. At the same time some of the earlier devices have now fallen into general disuse, but I have retained a description of them in order to maintain a historical perspective.

While revising the book I also took the opportunity to rearrange the earlier chapters to eliminate some of the repetition which crept into the first edition. I would still recommend that the first three chapters are not skipped as some of the examples and techniques used later on are developed at this stage.

An innovation is the inclusion of examples (that is questions) in the first five chapters. I hope that these will exercise the reader and help to assess the level of understanding achieved. From Chapter 6 onwards, examples become difficult to set as design software is needed to cope with the architectures and principles described.

The appendices are intended to provide references for the use of PLDs in practical situations; that is to point the reader to the suppliers of devices, and design and programming support. I have done my best to compile accurate and comprehensive lists, but I can take no responsibility for loss caused by errors or omissions.

There is no doubt that programmable logic is an established design resource and will continue to expand in scope and depth. This book should appeal to the student and experienced user; I know that I have often referred to my own copy of the first edition when confronted with a design problem. If every reader finds it as useful as I do then it will have been worth writing.

Geoff Bostock
1992

Preface to the First Edition

Programmable logic devices have been available for over 15 years yet, until now, there has been no comprehensive reference work covering the subject. This is because it has been possible, until recently, for the design engineer to choose from a fairly restricted range of devices; restricted both in the sense of being small in number and simple to understand. In the last few years, however, the choice of architecture has become wider and more powerful in its application possibilities.

In compiling the data for this book it was clear that the starting point was hard to define; how much previous knowledge of semiconductors and electronics could I assume that the reader has? The easy way out was to assume that the answer was none. The early chapters are a summary of the basics of semiconductor devices and the principles of logic but, because this is intended as a reference work, there are no formal proofs of the results obtained. The references at the end of the book contain any formal working required by the reader. However tempting it is to miss out these early chapters, which may appear trivial to readers well versed in electronics, some of the later points have their origins in these basic teachings.

To put programmable logic into perspective let us take a brief historical look at the subject. PROMs were the first devices to come to the market, in about 1970, although whether they were intended as memories or logic is a moot point. The first true logic device was Signetics FPLA, introduced in 1974. Initially this was a limited success for a number of reasons; high complexity (for that time), inflexible architecture, large package and high price (compared with standard logic) being the chief ones. PALs, introduced by MMI in 1977, overcame many of the drawbacks of PLAs and took a lead in market share which they have never relinquished. The total market size is now over a billion dollars, shared between more than 100 device types.

Programmable logic has become a real competitor to standard logic on the one hand, and to masked ASICs on the other, but it still represents only a small fraction of their market size. There is thus a huge potential for growth with consequent benefits to designers, provided that they are in a position to take advantage of those benefits. This book is intended to help them take fullest advantage. All the currently available architectures are described in detail; the design methods are also covered. These range from manual techniques to high-powered CAE systems, showing that programmable logic is suitable for both low budget projects and highly equipped design centres.

One of the largest sections of the book is devoted to applications. These include simple logic functions which can be used to build more complex functions and some ideas for the more complex functions themselves. The applications are intended to be diverse enough to show how most logic requirements can be fitted into programmable logic devices. Even if the application is not covered exactly, enough information is provided to enable the designer to make the best choice of device and be guided as to how to complete the design.

The student with no prior knowledge of programmable devices should find this work a useful primer, particularly when used in conjunction with standard text-books covering the formal aspects of logic design. I hope that established designers will also find it helpful as a reference work to keep by their benches. If it helps to breed a generation of logic designers who turn to programmable devices before looking at lists of TTL and CMOS standard functions, then it will have achieved its goal.

Geoff Bostock

Chapter 1
Introduction to Logic Devices

1.1 BASIC PRINCIPLES

1.1.1 The idea of logic

1.1.1.1 AND function

In order to understand any discussion of what comprises a logic device it is necessary to be aware of what is meant by logic. In any system of logic, be it electrical or philosophical, the fundamental concept is that statements may be *true* or *false*. Conclusions about the state of the system being described are drawn from an analysis of which components of that system are true and which are false. For example, a simple combination lock might be devised in which two-way switches are placed in series with a relay. The relay would operate the lock mechanism; the system circuit is shown in Figure 1.1.

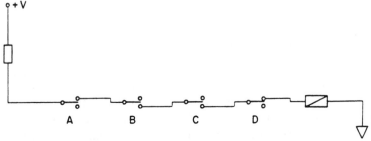

Fig. 1.1 Combination lock with switches.

Each switch is given a letter and only one combination of ups and downs will allow the lock to operate. In the system illustrated, with only four switches, it would not be very difficult to break the combination for there are only sixteen possibilities, but we will see in later chapters how the number can be extended to make a practical circuit. In our simple system it can be seen that the combination of A-up B-down C-down and D-up will open the lock. If up is equivalent to the *true* state and down is the *false* state, then a *logic equation* may be written to describe the circuit:

OPEN = A AND NOT B AND NOT C AND D

AND and NOT are the *logic operators* which define the relationships between the variables. In this example A and D must both be true and B and C false for the equation to be satisfied.

The AND operation implies that all the stated conditions must be satisfied for the result to be true. Thus, in our example, every switch must be in its correct position for the lock to open. If the combination required all the switches to be up (true) then we could write the logic equation using the symbol '*' to represent the AND operation. Sometimes the symbol '.' is used for AND but most design software packages use the '*' so we will standardize on this. The equation becomes:

OPEN = A * B * C * D

NOT is the operation which changes true to false or vice versa. It is written in logic equations by putting a line over the negated symbol, or by placing a '/' before the symbol. Thus the original equation becomes:

OPEN = A * /B * /C * D

1.1.1.2   OR function

The third of the operators is the OR function, which provides for logical alternatives. To illustrate this let us add a second combination to the lock, perhaps so that two people could have their own access to the safe! Let the second combination be A-up B-down C-up and D-down. The logic must allow either combination to OPEN the lock so we use the OR function to combine the two combinations. The symbol for OR is ' + ' so we can write the equation as:

OPEN = A * /B * /C * D
       + A * /B * C * /D

Figure 1.2 shows how this logic system could be built from mechanical switches. The two-way switches used above are replaced by double-pole two-way switches and if either path is made the lock will open.

A logic system may be employed to make decisions about almost any situation. While the human brain is quite capable of making these decisions it usually needs the data to be converted to a visual form. Electronic logic devices take the data in the form of electrical signals and use electronic

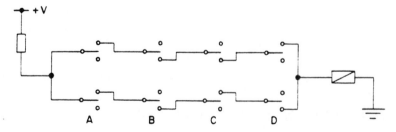

Fig. 1.2 Combination lock with switches – version 2.

switches to implement the logic equations. In our example we used mechanical switches as the logic elements as well as the interface between the outside world and the electrical system. Usually the interface is separated from the logic elements and from now on we will concentrate on the electronic devices used to perform the logic. The most common electronic switch is the transistor and the next section describes the two types used in practical circuits.

1.1.2 Transistor switches

1.1.2.1 Semiconductors

Before describing transistor operation it is necessary to appreciate the materials from which they are made. Matter under normal conditions is composed of atoms. In solid matter the atoms are bonded together and held in relatively fixed positions by the interaction of the electrons in their outer layers. Conducting materials, such as metals, do not use all their electrons for bonding so the spares are free to move within the solid boundary and will conduct electricity. Other substances, particularly those with complex molecules, have no spare electrons and are therefore insulators. The effect of temperature is also relevant.

When heated a solid absorbs energy; internally this energy is stored as vibration energy by the atoms or molecules. In a conductor this has the effect of reducing the available space for the electrons to move around in, so the bulk resistivity of the material is increased. The effect on insulators is different. Some of the energy is transferred to the electrons, which are then able to escape from their bonding duties and become free to conduct electricity. Those materials in which this property is noticeable at room temperature, particularly monatomic crystalline solids such as silicon and germanium, are called semiconductors.

There is a more controllable mechanism by which semiconductors may be made to conduct electricity. A small amount of an impurity may be added to a crystal without disturbing the lattice too greatly, provided that the atoms of the impurity are similar in size to the parent atoms. If the impurity has more electrons than the parent available for bonding the spare electrons become available for conduction. This is called an *n-type* semiconductor because the current is carried by negative charges. Conversely, it is possible that an impurity will have fewer electrons available for bonding than the parent, in which case there will be *holes* formed in the bonding layer. Under the influence of an electric field, electrons will move to fill adjacent holes leaving a hole where they were; this makes it appear as if the holes themselves are moving through the crystal. Such material is called a *p-type* semiconductor as the current is carried by positive charges.

1.1.2.2 Diode junctions

While the current-carrying potential of a simple semiconductor depends simply on whether there is an excess of electrons or holes, a quite different

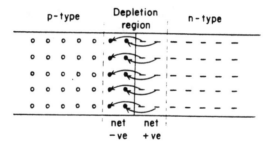

Fig. 1.3 p–n junction.

situation exists when there is a junction between p-type and n-type material. At the junction itself there is a thin layer called the *depletion region* where the material is *intrinsic* and there are no free charge carriers. Figure 1.3 shows how the free electrons from the n-type side can diffuse to the other side and fill the holes, creating a potential barrier.

Applying a positive voltage to the n-type side pulls electrons away from the depletion region and increases the height of the barrier. In this case no current can flow through the junction. In the reverse case, when a negative voltage is applied to the n-type material, electrons are repelled towards the depletion region and cause the potential barrier to be lowered. When the barrier has been eliminated they meet holes which have been attracted from the p-type side of the junction. The electrons combine with the holes allowing a continuous flow of electrons in one direction and holes in the other. The net result is that there is a constant current flowing through the junction.

This property of the p-n junction, allowing current to flow in one direction but not the other, forms the basis of most electronic components, from the diode to the VLSI integrated circuit.

1.1.2.3 MOS transistors

The MOS transistor is shown in cross-section in Figure 1.4. MOS is an acronym for metal-oxide-silicon which describes the basic structure. The transistor is fabricated from a crystal of p-type silicon, into which impurities are diffused to form n-type regions called *sources* and *drains*. A thin layer of silicon dioxide is grown above the gap between each source and drain and a layer of metal or silicon deposited on the top. This top layer is called the *gate* and controls the current flow between the source and drain.

If the source and bulk silicon are held at the same voltage and the drain is taken to a more positive voltage then no current can flow between source and drain because the drain-substrate junction is reverse-biased. If a positive voltage is now applied to the gate, electrons will be attracted into the region immediately below the oxide. This has the effect of making an n-type *channel*, which allows electrons to flow from source to drain. The voltage on the gate thus controls the flow of current through the transistor. A transistor of this

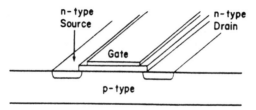

Fig. 1.4 MOS transistor structure.

type is called an *n-channel* device. A similar transistor made on an n-type substrate would be a *p-channel* device.

An MOS transistor can be used as a switch by connecting the current to be switched to source and drain, and connecting the controlling voltage to the gate. Because silicon dioxide is a good insulator, very little current has to be supplied by the control voltage; however, because the gate is acting as a capacitor there may be loading effects at high frequency. The channel is confined to a shallow region just below the surface and will therefore not permit very high currents to flow. The full consequences of these properties will be examined in a later chapter.

1.1.2.4 Bipolar transistors

If the n-type side of a p–n junction is doped more heavily with impurities than the p-type side, then many more electrons than holes will be attracted to the depletion region when the junction is *forward-biased*. Most of the electrons will then pass into the p-type material, where they will be 'eaten' gradually by holes. The electrons are said to be injected into the p-region. In a bipolar transistor, Figure 1.5, the n-type region is called the *emitter* and the p-type region the *base*. The base is made very narrow and bordered by a second n-type region, the *collector*, which is usually made more positive than the base. Most of the electrons injected into the base will be attracted into the collector thus establishing a current flow between collector and emitter.

The base controls the collector current, as the voltage between base and emitter determines how much electron current is injected by the emitter. If the base voltage is too low to allow injection then no current will flow in the

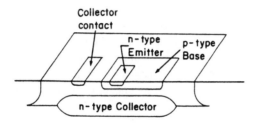

Fig. 1.5 Bipolar transistor structure.

transistor; thus the base acts as a control terminal for the bipolar transistor just as the gate does for the MOS device. Unlike the gate, the base must supply a small current to account for those electrons which are injected by the emitter but combine with holes before they reach the collector. However, because the area of the emitter can be made relatively large, a bipolar transistor can carry a larger current than an MOS.

The transistor described above is called an *npn* transistor to show the doping types of the three regions making up the device. It is also possible to make a *pnp* transistor where the dopings and voltages are reversed.

1.2   PRACTICAL LOGIC DEVICES

1.2.1   Planar technology

1.2.1.1   Masking and diffusion

Before embarking on a study of both standard logic families and programmable devices it is instructive to examine the technology used to fabricate integrated circuits. This is still based on the *planar process* developed in the late 1950s by Fairchild Camera and Instrument Corporation. Figure 1.6 illustrates the steps required to create a p-type region in an n-type crystal of silicon. Slices, or wafers, about 1 mm thick are cut from a silicon crystal which has been grown by the *Czochralski* method. Current production uses wafers up to 150 mm in diameter. The wafers are chemically polished to remove mechanical damage incurred in the cutting process.

The first step is to grow a layer of silicon dioxide on the surface of the wafer. This is achieved by passing oxygen over the wafer in a furnace at a temperature of up to 1200° C. As many as 50 wafers may be processed at one time and furnace temperatures are controlled to better than 1° C. A thin layer of sensitive material is then applied to the surface of the wafer and the areas where p-type regions are required are defined. This may be achieved by exposure to ultraviolet light or an electron beam which polymerise the layer in areas which are to remain n-type. A photo-mask is used with ultraviolet light; the electron beam, which gives much finer definition, is electrically scanned over the wafer. The unpolymerised areas are dissolved in solvent to reveal the silicon dioxide surface, which is then chemically etched by hydrofluoric acid to expose the underlying silicon.

The next step involves another high temperature furnace operation; in this case a gas containing the required impurity is passed over the wafers and forms a solid solution at the surface of the silicon. Alternatively, the wafer can be made the charged target of an *ion implantation* process. Charged atoms, or ions, are attracted to unmasked areas of silicon and become embedded in the surface layer. This process gives a more accurate dose of impurity than vapour deposition, but it does result in damage to the crystal lattice which must be annealed out at the next stage.

Prolonged exposure to temperature then causes the impurity to diffuse into

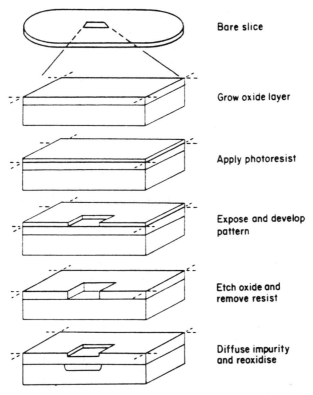

Bare slice

Grow oxide layer

Apply photoresist

Expose and develop pattern

Etch oxide and remove resist

Diffuse impurity and reoxidise

Fig. 1.6 Masking and diffusion steps (planar process).

the silicon to a depth of a micrometre or more, and provides the necessary annealing to ion implanted material. The net result is a tub of p-type material in the n-type; the boundary between the two being a p–n junction. Successive diffusion steps are required to build up the transistor structures described in the previous section.

1.2.1.2 Metallisation

Having fabricated silicon based components a way has to be found to enable them to be used. This involves connecting some kind of rigid metallic structure to the silicon to enable the device to be mounted, for example, onto a printed circuit board. The size of transistor features, usually a few micrometres, means that direct connection of wires is virtually impossible for reliable permanent joints. In practice photo-lithography has to be used again. A metal, such as aluminium, is evaporated over the silicon surface to form a thin film. Windows previously etched in the silicon dioxide allow the aluminium to make contact to the silicon where connection is required. The aluminium itself is then etched to form conductive tracks from the silicon windows to metallic areas which are large enough to allow direct connection of wires.

Where the aluminium contacts the silicon an alloy is formed to ensure reliable connection. Aluminium is a p-type impurity in silicon so some care has to be taken when connecting to n-type areas. If the n-type area is doped heavily enough the depletion region will be extremely thin and holes and electrons will cross it very easily even when it is reverse biased. There will be no problem in connecting aluminium to p-type silicon as no junction is formed in this case.

1.2.1.3   Integrated circuits

As well as providing a conducting path from the silicon to the connecting wires, aluminium tracks can be used to connect diffused components in the same silicon wafer. Circuits containing several transistors, diodes, resistors and even capacitors can be connected up on the silicon wafer surface, just as discrete components can be connected on a printed circuit board. In principle there is no limit, within the ingenuity of the designer, to the number of components which can be connected on a wafer. In practice the planar process is subject to random faults caused by dirt particles or material defects. If a component has a fault it will not function correctly, so the circuit containing it will also be defective.

The potential number of circuits on a wafer depends on the area of the wafer and the area of a circuit. Probability theory can predict what proportion of these circuits will be faulty for a given level of fault densities. This enables manufacturers to calculate the largest circuit which they can make economically. Improvements in processing and materials technology cause a steady reduction in the fault density allowing larger circuits to be designed and manufactured.

1.2.1.4   Packaging

The final stage in integrated circuit manufacture is packaging. Most electronic assemblies are based on printed circuit technology so the integrated circuits have to be packed into a form which is robust enough to withstand the handling they will receive, while allowing the connecting pins to be soldered into the *PCB*. The standard package is the *dual-in-line* which has two rows of pins fitting into a 0.1 in. grid. The pins are 0.1 in. apart; the row separation depends on the number of pins, usually 0.3 in., 0.6 in. or 0.9 in.

The packages are constructed on a frame, called the *lead frame*; the circuit itself is alloyed to a central bar and fine wires, about 25 micrometre thick, bonded from the circuit to the pins. The whole frame is either sandwiched between two ceramic plates in an inert atmosphere, or moulded into solid plastic. The circuit is cropped out of the supporting frame and the leads formed into the conventional inverted 'U' to allow easy insertion into holes in the printed circuit board. Package sizes from 8 to 64 pins are found in this form; the number of pins and package width are related by the mechanical problems of fitting the leads round the circuit. Reducing package size for a given pin-count is a major task, which has received the attention of most i.c. manufacturers.

A recent development is the *SO package* which has leads on a 0.05 in. spacing formed into an 'L' shape; these allow the circuit to sit on the board, instead of the pins passing through holes. The package may be stuck to the board and connection made by means of a solder paste laid down previously by a screen printing process. The smaller spacing means smaller packages, and other components are available in similar styles. Component placing may be mechanised so the production process becomes cheaper, and the size reduction also means cheaper mechanical components. The whole process is called *surface mounting* and is being used by a significant number of equipment manufacturers.

As logic functions become more complex they often need more connections, and the DIL and SO packages cannot cope with the pin numbers required. Another type of surface mounting package has been developed, based on having leads on all four sides. The first packages of this kind had a ceramic base with printed connections leading to plated areas at the edge of the package. These were intended for direct soldering to a ceramic substrate, as found in thin and thick film circuits. Special sockets were developed to enable these packages to be used on printed circuit boards; their lack of leads led to them being called *leadless chip carriers*, LLCCs for short. Plastic versions in the same style use leads bent into a 'J' and located in notches at the edge of the package; they are called *plastic leaded chip carriers*, or PLCCs. PLCCs and LLCCs have been developed for packages as small as 20 leads, but they are more useful for circuits with much larger counts, from 44 to over 150 are now possible.

1.2.2 Practical logic circuits

1.2.2.1 Bipolar logic

Bipolar transistor action depends on the properties of p–n junctions which are below the silicon surface, albeit by only 1 or 2 micrometres. The active region of an MOS transistor is at the surface itself. The surface properties of silicon, and the ways to control them, were understood at a later time than the bulk properties so bipolar logic circuits evolved before MOS. Figure 1.7 shows how bipolar silicon components can be connected to form an AND circuit.

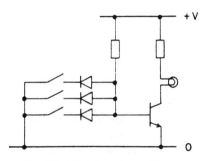

Fig. 1.7 AND function – bipolar.

If all three switches are open then no current will flow in any of the three input diodes. The base of the output transistor is thus connected to a positive voltage, so the transistor is switched on, current flows in the output resistor and the lamp will be turned on. If any of the switches is closed current will flow through the diode connected to it, so the voltage on the transistor base will be insufficient to allow it to conduct appreciably, and the lamp will not be lit. The lamp turns on only when all three switches are open so the circuit implements the AND function.

1.2.2.2   MOS logic

Once the properties of silicon surfaces were understood, and the techniques for controlling them had been perfected, it became possible to construct logic circuits using MOS switches. The circuit in Figure 1.8 shows how a simple NOR gate can be built from MOS transistors. The transistors are *n-channel* type so any gate taken to a positive voltage will open a conductive path from the load resistor to ground. This structure will, however, provide no particular advantage over bipolar transistor circuits; indeed, the performance is likely to be worse because of the lower current capability and higher capacitance of MOS.

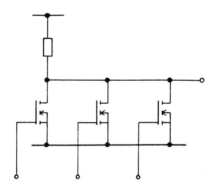

Fig. 1.8 NOR function n-channel MOS.

The identical function can be constructed from *p-channel* transistors, as shown in Figure 1.9, by connecting them in series. Any gate taken to a positive voltage will turn its channel off and prevent current from flowing in the chain. Again no advantage is obtained from this structure, but consider what is achieved by combining the two, as in Figure 1.10. If any inputs are taken positive the *p-channel* transistors are switched off, while if all the inputs are negative the *n-channel* transistors do not conduct. Thus no current flows in any steady state condition. The result is a logic circuit taking very little current; there is now a decided advantage over bipolar transistors.

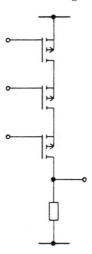

Fig. 1.9 NOR function – p-channel MOS.

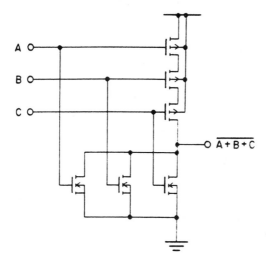

Fig. 1.10 NOR function – CMOS.

1.2.2.3 Technology comparison

Some of the advantages of bipolar against MOS and vice versa have already been mentioned above. The choice between the two technologies is essentially dictated by practical considerations which are worth summarising here. The two chief performance criteria under consideration are usually speed and power. That is how fast the circuit will perform the function contained in it, and how much power must be supplied to it in operation. Usually the faster the circuit the higher the power consumption because speed is achieved by charging circuit capacitance quickly, which involves the use of higher electrical currents.

As already noted, bipolar transistors are better at handling high currents and so bipolar technologies usually offer higher speed. Examination of the technology shows, however, that MOS components can be diffused closer together and intrinsically smaller. This is because bipolar transistors need to be isolated from each other while groups of MOS can be diffused together. Also the diffusion windows in bipolars need aligning with each other while many of the MOS processes are self-aligning. MOS structures therefore have lower capacitance, and can operate at lower currents for a given speed. This is no real advantage where simple circuits are concerned because the major component of capacitance is in the external circuit which is connected to it. As circuits become more complex the internal capacitance becomes more important and MOS becomes a better choice for high speed.

The MOS circuit described above using both p and n-channel transistors is termed *complementary MOS*, or CMOS for short. The other main advantages of CMOS are the higher noise immunity inherent in the gate design – CMOS switches virtually between the two power rail voltage levels – and more complex functions can be built into the same area of silicon because of the higher packing density of MOS transistors. On the bipolar side is the possibility to make the absolutely fastest circuits, at the cost of high power consumption, and greater robustness, because MOS transistors are susceptible to relatively low static electricity voltages on their input gate oxide.

A physical device performing a simple logic operation is also called a *gate* because of the analogy between an electrical switch and a gate, which allows entry when open. Gates are formed from either bipolar or MOS transistors, as we have just seen, and we will now show how these are made the basis of standard logic families.

1.2.3 Standard logic families

1.2.3.1 TTL

One of the prime considerations of a family of logic devices is that they must be capable of being connected together. That is, they all recognise the same voltages as logic HIGH and logic LOW, and when the output of one is connected to the input of another there is not a major change in output voltage because of the current flowing between the two. A number of bipolar families were introduced when integrated circuits were in their infancy but the transistor–transistor–logic or TTL family was soon established as a *de facto* standard.

Figure 1.11 shows the basic TTL NAND gate. The gating takes place in the multi-emitter input transistor, which is a compact development of the standard input diode cluster. The signal is shifted back to standard levels by

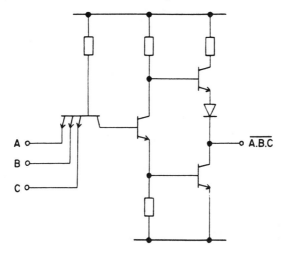

Fig. 1.11 NAND gate – TTL.

the rest of the circuit, which also provides sufficient output current to drive succeeding logic stages. The standard TTL voltages are:

supply voltage– 5.0 V
HIGH voltage– >2.0 V
LOW voltage – <0.8 V

Usually there is sufficient output current to drive ten other TTL circuits, a property called *fan-out*.

One problem with TTL is that the internal transistors are 'overdriven' to ensure that they are fully switched on. The extra drive current causes the base region to act as if it were a capacitor; when the transistor is switched off this capacitor has to discharge through a reverse-biased diode junction. This 'saturation' delays the switch-off and slows down the reponse of the circuit.

The original TTL family used 'gold-doping' to reduce the effect of this problem. Gold creates 'traps' in the bulk silicon; these soak up the excess charge and speed up the switch-off.

A later, and better, solution diverted the excess base current through a *Schottky* diode in parallel with the base–collector junction. A Schottky diode is a metal-silicon diode which has a much lower capacitance than a diffused junction diode. It forms the basis of the high speed STTL and LSTTL families.

1.2.3.2 ECL

Although it is possible to make OR gates in the TTL families, TTL is essentially NAND based logic. A family of circuits based on the OR structure described earlier has also been developed. This is the emitter-coupled logic or ECL family. It has been refined by using a comparator principle to determine whether input voltages are HIGH or LOW. The emitter-coupled pair of

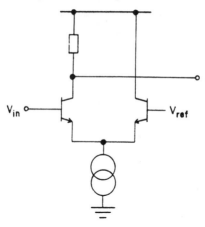

Fig. 1.12 Emitter-coupled-pair comparator.

transistors in Figure 1.12 form a simple comparator such that current will flow in the load if the input voltage is above the reference voltage, but not if it is below the reference. By connecting the input transistors in parallel a NOR gate can be constructed.

A complete ECL gate includes a level-shifting emitter follower, as shown in Figure 1.13, to give the correct voltage to drive the next stage. Because it uses higher current levels and a lower voltage swing, but chiefly because it is designed to be non-saturating, ECL gates have a shorter delay than TTL; ECL is to be found more often in higher speed applications. Unlike most other logic families, it is powered by a negative supply; that is the most positive voltage is 'ground', because the voltage swing across the load is more accurately defined by reference to a positive ground. The standard voltage levels are:

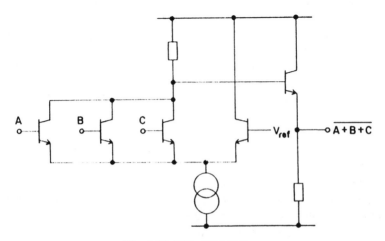

Fig. 1.13 ECL OR/NOR gate.

supply voltage – $-4.5\,V$ or $-5.2\,V$
HIGH voltage – $>-0.8\,V$
LOW voltage – $<-1.6\,V$

Fan-out is quite high, but more important are the line driving abilities of the outputs in these high speed applications.

1.2.3.3 CMOS

The basic CMOS NOR gate has already been described in section 1.2.2.2. By inverting the series–parallel arrangement of the MOS transistors, the CMOS NAND gate of Figure 1.14 can be made.

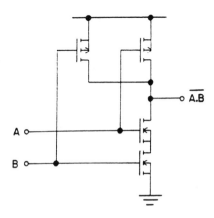

Fig. 1.14 CMOS NAND gate.

The limitations of CMOS gates are somewhat different from the bipolar logic families. The chief disadvantage is the restriction on the number of inputs to a simple gate. When turned on, an MOS transistor has a relatively high series resistance, so a load formed by transistors in series will show a high output resistance. On the other hand, the resistance of parallel transistors depends on how many of them are turned on. The latter can lead to disturbing changes in output resistance as input signals to the gate change sense. The former, however, can prevent the gate from working at all. If a CMOS NAND gate is sinking too much current the output voltage may rise to such a level that the input voltage to the transistor on top of the string may not be enough to turn it on. Moreover, the threshold voltage in absolute terms will be different for each input.

The standard solution to these problems is to buffer the outputs with two inverters; this does not affect the logic function but does prevent conditions at the output either influencing, or being influenced by, conditions at the inputs.

CMOS does have a big advantage over bipolar circuits in that it is very tolerant to supply voltage variations. Provided that the voltage swing is sufficient to turn the transistors on and off, 'overdriving' them has no effect provided that the breakdown voltages are not exceeded. Thus CMOS voltage levels are defined somewhat differently from bipolar, as follows:

supply voltage – 3 V to 15 V
HIGH voltage – $>0.7 \times$ supply voltage
LOW voltage – $<0.3 \times$ supply voltage

The fan-out of CMOS gates is limited more by a.c. considerations than by d.c., because the input to a CMOS gate looks like an almost pure capacitance. Driving several inputs delays signals, but makes virtually no change to their voltage level.

Although the traditional CMOS '4000' series can use the wide voltage range above, the new '74HC' family operates at TTL supply voltages and the '74HCT' version interfaces directly to TTL. Also the more complex CMOS circuits, such as PROMs and microprocessors, normally interface with TTL voltage levels.

1.2.4   Gate symbols

When drawing logic systems standard symbols are used to represent the logic functions. There are now two systems of symbols; the traditional symbols, which we will use, and a new system devised by the IEEE/IEC. Figure 1.15 shows the symbols for the gate functions described so far. It should be noted that inversions are shown by a 'bubble' on the output of traditional symbols. An inverting bubble may also be placed on inputs when an input function is being inverted.

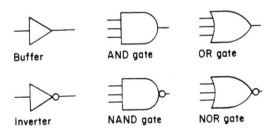

Buffer	AND gate	OR gate
Inverter	NAND gate	NOR gate

Fig. 1.15 Standard gate symbols.

1.3   CUSTOM LOGIC CIRCUITS

1.3.1   Microprocessors

The first integrated circuits were only simple functions capable of containing a single logic equation. Attempts to make larger circuits failed because the silicon processing introduced random faults, which limited the area of silicon usable by a single circuit. This was acceptable to equipment designers, who were used to thinking in terms of what could be built from discrete transistors on a single printed circuit card. The replacement of a dozen components by a single metal can containing the same components in miniature form was an obvious benefit to them. Improvements in silicon processing meant that more complex circuits could be integrated, but only up to a point. There are certain logic functions which are used universally in system design and it clearly makes sense to integrate these, but above this point the way that these functions are combined becomes specialised or *application specific*.

One way in which the component industry reacted to this situation was a result of its close involvement with the computer industry. Computer manufacturers were the strongest influence on component designs because the performance of computers was strongly dependent on the speed and complexity of the circuits from which they were built. A computer is a highly complex logic system capable of implementing logic and arithmetic equations by virtue of a set of instructions which modify the internal logic functions according to the instruction being carried out. There is no reason in principle why a computer cannot be built on a single silicon circuit and this was the route taken by some component manufacturers. Because of their small size, and because the number of functions they could manage was limited, the name *microprocessor* was coined for these devices. The function of the micro-processor was determined by the sequence of instructions they were controlled by, rather than their actual circuit layout, so they were complex circuits which could be made application specific by the equipment designer.

1.3.2 Custom circuits

The main reason for the upper limit on complexity of standard integrated circuit functions is concerned with the economics of integrated circuit manufacture. The cost of designing even a simple circuit and getting it into production is more than £100 000 ($100 000), so manufacturers rely on selling millions of circuits to recover the design costs. In some cases, however, an equipment manufacturer may feel that it is worthwhile to invest that sum to obtain the benefits of having a customised circuit built for them. The equipment maker will save on his production costs by replacing several circuits with one and will have the added advantage of secrecy, since it would be difficult for a competitor to copy his system without knowledge of the custom circuit.

This course of action will only be possible if a substantial number of customised units can be built – probably well in excess of 10 000. Integrated circuit manufacture is a batch process in which many thousands of good circuits should result from a single batch, even for complex designs. Because of the other setting up costs, such as mask making and test program writing, most manufacturers would be unwilling to commit these resources to just a single batch of custom circuits. The full-custom option then is only viable for well proven designs committed to long production runs. Another problem for the unwary is that it is not always straightforward to transfer a design involving several discrete circuits onto a single silicon circuit. Typical problems are caused by timing changes and layout interactions which may result in several iterations of the mask set. Good computer simulation of the circuit can cut down the problems; however, a job which may start out with an expectation of completion in 12 months, long enough for a circuit design anyway, may finish up taking twice as long and costing twice as much.

1.3.3 Gate arrays

Many manufacturers have introduced the *gate array* concept which is a good compromise between standard and custom circuits. A gate array has a base layer containing hundreds or even thousands of identical simple logic functions diffused in a regular pattern, or array, in the silicon. The whole design is usually very software intensive with a library of standard functions called *macros* which can be built from the basic functions. The designer defines the way in which the macros themselves are interconnected, and computer programs then physically define the layout on the array and simulate the electrical performance based on this layout. Iteration to obtain the desired performance is relatively quick and the program will usually produce the interconnection masks and test programs automatically. It now becomes economic to produce only a thousand circuits of a given design and allows smaller users to enjoy the benefits of customisation.

Being a compromise this approach does not usually yield as high a performance as full-custom solutions, nor is it as flexible in the range of circuit complexities which can be accommodated. Other approaches use a variation of the gate array topology by offering, for example, standard cells, completely flexible gate counts or direct writing by electron beam to improve performance or speed-up prototype deliveries. In all these cases there are a number of common factors:

- a commitment to some minimum quantity
- a development charge payable to the manufacturer
- the manufacturer controls the final production

Thus the consumer who wants just a few, or is unwilling to commit up front to a full design, or needs a fast delivery response is unlikely to be an array or custom circuit user.

1.4 EXAMPLES

1.1(a) If the resistor values in Figure 1.11 are 4.0 kohms (input transistor base resistor), 1.6 kohms ('middle' transistor collector resistor) and 1.0 kohms ('middle' transistor emitter resistor), what are the internal node voltages and output voltages when all inputs are at 2.4 V?

(b) What are the voltages if one input is taken to 0.4 V?

(c) What is the function of the collector resistor of the upper output transistor?

(Assume the forward diode voltage, V_f or V_{be}, is 0.7 V and the saturation voltage, V_{sat}, is 0.3 V)

1.2(a) In the CMOS NOR gate of Figure 1.10, what will be the effect on output voltage if, respectively, one, two and three inputs is HIGH? Assume that the output is sinking 10 mA and that the on resistance of an MOS transistor is 150 ohms.

(b) If the gate is operating at 5 V and the transistors have a threshold voltage of 1.5 V, what minimum voltage will be required at each input to take the output HIGH if both other inputs are at 0 V?

Chapter 2
More Logic Functions

2.1 REPRESENTATION OF LOGIC

2.1.1 Truth tables

In Chapter 1 we saw how to represent logic functions, such as AND gates, by equations which define an output variable in terms of input variables. This may not always be the easiest way to visualise what is required from the logic function being described. For example, we wrote an equation to describe the action of a combination lock in Section 1.1.1.2, and we showed how this could be built from switches. Anybody wanting to operate the lock must be given the combination in some easily remembered format. The equation is not the most memorable form to retain this information. Much easier would be to describe the two combinations as 'ON, OFF, OFF, ON' and 'ON, OFF, ON, OFF', both of which have some pattern to them. This is the basis of the truth table.

As its name implies, it is a table of true conditions. In electronic logic systems we usually refer to 'HIGHs' and 'LOWs', rather than ON and OFF, and use the abbreviations 'H' and 'L' to refer to them in truth tables. The truth table representation of the combination lock is shown below:

A	B	C	D	OPEN
				H *(active level)*
H	L	L	H	A
H	L	H	L	A

There are, of course, 16 possible combinations of H and L for the four inputs but only the two true, or active, combinations have been listed, it being understood that the others are inactive. The active output level has been defined as 'H' so it is implied that all other combinations will give a LOW output.

In some cases all the combinations are defined for the sake of clarity or completeness but then, to save space, the symbol '×' or '−' is used to represent an input where either 'H' or 'L' will have the same effect (known as 'don't care'). An example of this is the complete truth table for an OR gate, shown below:

A	B	C	Y
H	–	–	H
–	H	–	H
–	–	H	H
L	L	L	L

It is implicit in the OR function that if input A is HIGH then the output will be HIGH no matter what the condition of the other inputs; but the output will only be LOW if all inputs are LOW.

2.1.2 Karnaugh maps

2.1.2.1 *Description*

Another way of representing a logic function, and providing a very useful tool for manipulating logic, is the Karnaugh map. A blank Karnaugh map for four input variables is shown in Figure 2.1. First of all it should be noted that the map contains 16 squares, that is one for each possible combination of the input variables. Secondly, it should be seen that the four variables are split into two groups of two, and each row or column of the map corresponds to a unique combination of the two variables. Thus the combination corresponding to each square is defined by the combination of the variables in the row and column meeting in that square. The square containing the asterisk is defined by A–'H' B–'L' C–'L' D–'H'. What is not so obvious is the way in which the rows and columns are arranged. The sequence of the combinations of variables is 'LL' 'LH' 'HH' 'HL', which is a method of counting known as *Gray code*. The key to counting in Gray code is that only one variable changes at any one time, and no combination is repeated. A Karnaugh map can be set up, in principle, for any number of input variables by writing the Gray code along the row and column axes and entering the output condition for each combination of input variables in the appropriate square in the map. Naturally this will be rather tedious for more than eight or ten variables although there are computer programs capable of handling larger maps.

A	0	0	1	1
B	0	1	1	0

D	C				
0	0				
0	1				*
1	1				
1	0				

Fig. 2.1 Karnaugh map for four variables.

2.1.2.2   Using Karnaugh maps

Having seen how to create a Karnaugh map for a logic function we will now look at some of the ways in which they can be used. Figure 2.2 shows the Karnaugh map for our combination lock. The two squares with 'H's represent the two AND gates in the circuit diagram; in fact each square can be physically implemented by means of an AND or NAND gate, depending on whether it contains an 'H' or 'L'. Thus it shows immediately how a function may be constructed from simple gates. It may also be used to improve the performance of a design.

	A	0	0	1	1
	B	0	1	1	0
D	C				
0	0				
0	1				H
1	1				
1	0				H

Fig. 2.2 Karnaugh map – combination lock.

In one of the fastest TTL families, 74FAST, NAND gates have a maximum delay time of about 5.5 ns, while the figure for AND gates is about 6.5 ns. Thus the combination lock could be made about 1 ns faster by using 14 NAND gates driving an AND gate instead of two AND gates driving an OR gate. The AND and OR gates have the same speed but AND is needed instead of OR because we are looking for an absence of LOWs rather than a HIGH present. This might not be very efficient in terms of the number of circuits used by the design but it might be a crucial factor in some designs which are critical on speed. The Karnaugh map may help even further.

If we look at the top right corner of Figure 2.3 we can see that there are two 'L's in adjacent squares. These correspond to the functions A*B*/C*/D and A*/B*/C*/D, which implies that the variable B can be either HIGH or LOW without affecting the output. These two functions can therefore be replaced by the single function A*/C*/D. This is an example of a general principle which can be applied to Karnaugh maps called *logic minimisation*. Any pair of adjacent squares containing identical outputs may be combined into a single function, and this principle may be extended to quartets, octets and so on indefinitely. Figure 2.3 shows the Karnaugh map with the 14 'L' functions combined into two octets and two quartets; that is four gates in all, which could make it worthwhile to achieve the improved performance.

A further point about logic minimisation can be illustrated here. The four gates could equally well be achieved by combining the two left-hand columns into an octet, the third column as a quartet and leaving the two 'L's in the

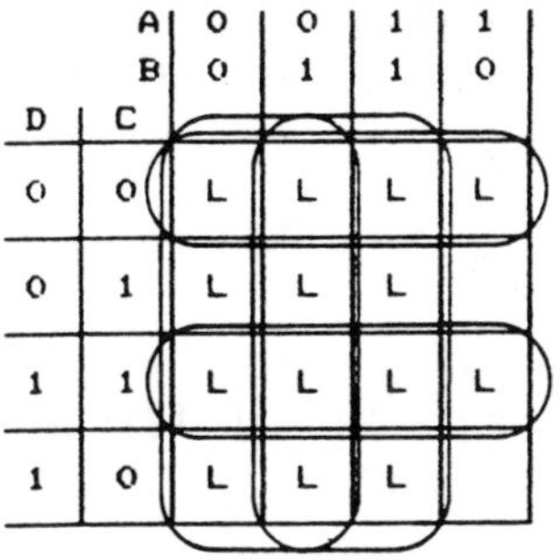

Fig. 2.3 Karnaugh map – combination lock ('L's grouped).

right-hand column as single functions. There are two reasons for proceeding as we did. Firstly, the two octets are physically implemented as inverters and the two quartets as 2-input gates, so the logic devices required are much simpler, leading to a more compact solution. Secondly, the alternative solution could give rise to spurious signals, known as *glitches*, when inputs change without causing an intentional change in the outputs.

Consider the two input combinations A∗B and A∗/B∗/C∗/D with A HIGH, C and D LOW, and B changing from LOW to HIGH. To start with only A∗/B∗/C∗/D is true so the output AND gate will see HIGHs from the other three NAND gates and a LOW from A∗/B∗/C∗/D, so it will have a LOW output. At the finish only A∗B is true so it will put a LOW onto the AND gate and its output will remain LOW. Now suppose that the delay through A∗B is longer than the delay through A∗/B∗/C∗/D, there will be a short period when all four NAND gates have a HIGH output so the output will go HIGH momentarily. This short glitch could be interpreted as a valid signal by following circuitry and cause a malfunction.

The solution which we have proposed overcomes this problem by ensuring that there are at least two NAND gates with LOW outputs either before or after a change in a single input. A glitch-free solution may always be ensured if groups of variables are overlapped wherever possible, as in this case.

2.1.3 Boolean algebra

We have already seen that logic systems may be represented by equations, as in our combination lock. Just as there is a set of algebraic rules for manipulating mathematical equations, there is also a set of rules for working with logic equations. They were first postulated by Boole, so the process of manipulating logic equations is commonly referred to as *Boolean algebra*.

In many ways Booles rules and theorems mirror those of mathematics. For example, the order in which variables are written is not significant, for

$$A∗B = B∗A$$
$$A + B = B + A$$

correspond exactly to their algebraic equivalents. The rules for expanding brackets, however, are not quite as straightforward:

A*(B+C)   =A*B+A*C

looks obvious enough but

(A+B)*(A+C)=A+B*C

needs some justification.

By drawing up a truth table to derive the LHS and RHS of the equation, we can show that they are identical:

A	B	C	A+B	A+C	LHS	B*C	RHS
0	0	0	0	0	0	0	0
0	0	1	0	1	0	0	0
0	1	0	1	0	0	0	0
0	1	1	1	1	1	1	1
1	0	0	1	1	1	0	1
1	0	1	1	1	1	0	1
1	1	0	1	1	1	0	1
1	1	1	1	1	1	1	1

The most powerful rules in Boolean algebra are known as DeMorgans laws and deal with the relationship between the AND, OR and NOT functions. They are as follows:

/(A+B+C+D+...)=/A*/B*/C*/D*...
/(A*B*C*D*...)   =/A+/B+/C+/D+...

Once again a truth table may be used to justify the laws, but this will be left as an exercise for the reader. However we can show these rules being applied to our combination lock equations to perform the same transformation that we achieved in the previous section with the Karnaugh map. We start with the equation:

OPEN=(A*/B*/C*D)+(A*/B*C*/D)
      =/(/A+B+C+/D)+/(/A+B+/C+D)   – DeMorgans first law
      =/(((/A+B+C+/D) * (/A+B+/C+D))   – DeMorgans second law
      =/(/A+B+(C+/D)*(/C+D))   – expanding brackets
      =/(/A+B+C*(/C+D)+/D*(/C+D))   – expanding brackets
      =/(/A+B+C*D+/C*/D)   – expanding brackets

These are just the four terms which we created by grouping 'L's in the Karnaugh map of the same function.

2.2 COMPLEX LOGIC FUNCTIONS

2.2.1 Combinational functions

2.2.1.1 Definition

In logic circuits we find that certain functions occur quite frequently in many varied applications. We can look at these standard functions and see how they fit in with the ideas which we have already established.

To start with we shall look at *combinational* functions, in which input signals are combined according to a logic requirement to give a predefined output signal. It does not matter in which order the signals are applied, or what the output state was before the inputs were changed, a certain combination of inputs will always lead to the same output condition. Multiplexers and decoders are examples of combinational logic functions.

2.2.1.2 Decoders and multiplexers

We have already seen an example of a decoder in our combination lock circuit, where a given combination of input signals will cause an AND gate to go HIGH if the LOW signals are inverted first. In general, any physical property of a system needs to be encoded before it can be processed by an electronic circuit. For example, the numbers 0–7 are usually encoded into binary 000, 001, 010, etc.; a decoder will transform each valid combination of the encoded input into a unique output signal.

A decoder, as in the combination lock, needs an AND gate for each output with every input signal, or its complement, being gated together. If there are n input signals then there are 2^n possible combinations of inputs. There must be 2^n AND gates in an n-input decoder. This is illustrated in Figure 2.4 which shows a 3-input, or 8-bit, decoder. The inputs are *buffered* so that each presents only a unit load to the driving circuits. There is also a fourth input which can be used as an *enable* line; by connecting it to every gate in the circuit it must be HIGH for the circuit to operate at all. This makes it possible to combine several small decoders into a large decoder.

The multiplexer in Figure 2.5 uses exactly the same principle as the decoder. The three address lines select one of the AND gates because only one combination of HIGHs and LOWs is true for each gate. A separate data input is the fourth input to each gate so the output of the selected gate reflects its data input. An enable line provides a fifth input to each gate. All the deselected gates have a LOW output so only the output from the selected gate can affect the output from the OR gate. The multiplexer output is thus the same as the selected input and is not affected by non-selected inputs.

2.2.1.3 The exclusive-OR function

The exclusive-OR function forms the basis of many arithmetic circuits and so merits a separate mention. Exclusive-OR could be called the inequality

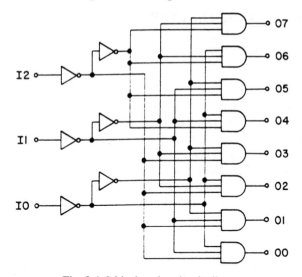

Fig. 2.4 8-bit decoder circuit diagram.

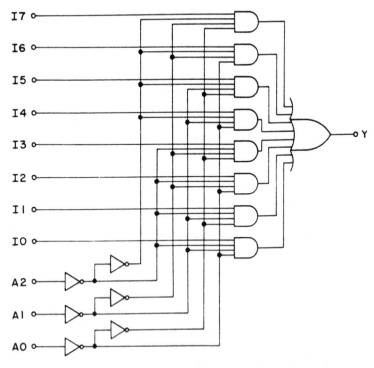

Fig. 2.5 8-input multiplexer circuit diagram.

function for it defines as being true when its two inputs are different from each other. In other words its truth table is:

A	B	$A: +:B$
L	L	L
L	H	H
H	L	H
H	H	L

The symbol ': $+$:' is one way of representing exclusive-OR in equations. If we look at the Karnaugh map for exclusive-OR in Figure 2.6 we can see that it may be built from AND and OR gates according to the equation:

$$A: +:B = /A*B + A*/B$$

Further examination of its truth table shows why it is important in arithmetic rather than pure logic circuits, although it frequently appears in those as well. A and B may be considered to be single digit binary numbers by calling the LOW state binary '0' and the HIGH state binary '1'. In this case A: $+$:B is the least significant bit of the arithmetic sum of A and B. Exclusive-OR gates are widely used in building full adders and other more complex arithmetic circuits.

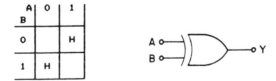

Fig. 2.6 Karnaugh map – 2-input exclusive-OR gate.

2.2.1.4 *Parity generators*

One use of the exclusive-OR which is worth dwelling on further is the parity generator. It is used to determine the number of HIGHs in a logic function, and an exclusive-OR gate is just a 2-bit parity generator. An even number of HIGHs in the function yields a HIGH output while an odd number yields a LOW output. Parity is used as a simple check on the integrity of data by seeing if the measured parity agrees with a parity bit associated with the data and calculated earlier.

We may try to build a parity generator from AND and OR gates by mapping the function onto a Karnaugh map, with a typical result as in Figure 2.7. The resulting 'checkerboard' pattern is the same however many inputs are considered; for the 6-input case illustrated, 32 AND gates and a 32-input OR gate would be needed. However, a much neater solution is obtained by using exclusive-OR gates as shown in Figure 2.8. The function can thus be built

			A 0	0	0	0	1	1	1	1
			B 0	0	1	1	1	1	0	0
D	E	C 0	1	1	0	0	1	1	0
D	E	F								
0	0	0	H		H		H		H	
0	0	1		H		H		H		H
0	1	1	H		H		H		H	
0	1	0		H		H		H		H
1	1	0	H		H		H		H	
1	1	1		H		H		H		H
1	0	1	H		H		H		H	
1	0	0		H		H		H		H

Fig. 2.7 Karnaugh map – 6-input parity generator.

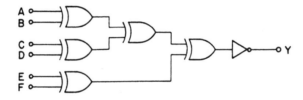

Fig. 2.8 6-input parity generator circuit diagram.

from just five exclusive-ORs and an inverter, less than half the hardware of our first attempt.

The lesson to be learnt from this example is that a single-minded approach does not always give the best result. In this case the high incidence of diagonal outputs would hint to an experienced logic designer that exclusive-OR might lead to a better solution than simple gates.

2.2.2 Sequential functions

2.2.2.1 Storage elements

Earlier in this chapter we defined a combinational function as one in which the outputs depended solely on the logic signals present at the inputs irrespective of the order in which they were applied, or the state of the outputs before they were applied. We can now consider a second type of logic device in which the output state does depend on the order in which signals are applied and on what the output state was previously. These devices are called *sequential functions* because, as their name implies, they depend on the sequence in which things are done to them. In order to achieve this they must have *storage elements* built into them so that they can 'remember' their immediate past history.

To illustrate this we can look at a typical microprocessor architecture. A microprocessor needs three types of input in order to receive instructions and communicate with the outside world. Each input consists of a group of signal lines, called a *bus*, which carry encoded data. The data bus carries information for processing to and from the outside world, the instruction bus carries the coded instructions for telling the microprocessor how to process the data, while the address bus is an output from the microprocessor which tells its peripheral circuits the source or destination of instructions or data.

In the simplest system these buses are separate, however, this type of structure leads to integrated circuits with a very high pin count so it is much more common for microprocessors to share pins for some functions; very often it is the data bus which is shared with some of the address lines. For part of the time the microprocessor is sending out the address of the next instruction while for the remainder it is using the pins as a data port. Fortunately for the user there will be a separate signal telling him what the pins are doing at any particular time. This means that when he is told that an address is being sent out he must store it, so that he can send the instruction back to the microprocessor at the right time. This more compact architecture is illustrated in Figure 2.9, together with a timing diagram showing the systems requirements; a ROM is a device we shall meet in Chapter 3, and is commonly used for storing instructions.

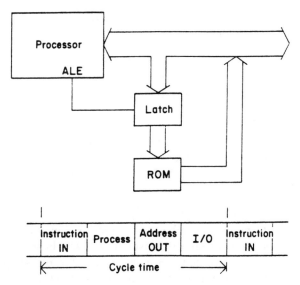

Fig. 2.9 Microprocessor architecture using mutiplexed address/data bus.

In this example we have assumed that the processor works in four 'phases'; in phase 1 it requires an instruction, in phase 2 it carries it out internally, in phase 3 it outputs the new address for the next instruction while phase 4 is used for data transfer to other parts of the system. During phase 3 a signal called ALE (address latch enable) goes HIGH to indicate that is the time when the address is being given out.

2.2.2.2 Basic D-latch

We will now see how to build a circuit element capable of storing one bit of the address. Its output must be HIGH when ALE is HIGH and when the input is HIGH; it must also be HIGH when ALE goes LOW and the output is already HIGH. This is achieved by feeding the output back as another input and gating it with /ALE. The logic equation is thus:

$$OUT = IN*ALE + OUT*/ALE$$

The output can change only when ALE is HIGH because when ALE is LOW the output is feeding itself; it is said to be *latched*. The usual symbols for input and output are 'D' (for data) and 'Q', and this type of latch is called a *D-latch*. If we look at the Karnaugh map for the D-latch, in Figure 2.10, we will see that

D	0	0	1	1
Q	0	1	1	0
LE				
0		H	H	
1			H	H

Fig. 2.10 Karnaugh map – D-latch.

the design is not quite complete. The two AND terms in the logic equation do not overlap so there is a risk of causing a glitch. In order to guard against this we must include a third term giving the complete equation as:

$$Q = D*LE + Q*/LE + D*Q$$

Note that this circuit fulfils the requirements of being a sequential circuit because Q can be either HIGH or LOW when LE is LOW and D is HIGH depending on whether D went HIGH before or after LE went LOW.

2.2.2.3 D-type flip-flop

One of the problems with combinational logic is the propagation delay through the logic elements, and we have already seen how this can lead to glitches if proper care is not taken with the design. One way to get round this problem is to only sample the logic functions at fixed points in time so that data moves through the system in discrete jumps, and it does not matter what

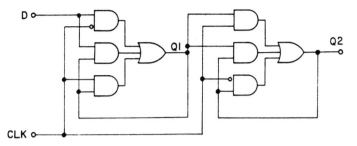

Fig. 2.11 D-type flip-flop circuit diagram.

happens between those jumps. The D-latch is not suitable for this purpose because data can move from input to output at any time when LE is HIGH. By putting two D-latches in series we can create the function which we require, as shown in Figure 2.11. If the first latch has an inverter in its enable input (i.e. /LE) then data can reach Q1 when LE is LOW. Once LE goes HIGH the data is passed to the output of the second latch, Q2, but Q1 is now fixed until LE goes LOW again. Thus data is transmitted through the two latches at the moment when the enable changes from LOW to HIGH.

The first latch is usually called the master and the second is called the slave, the whole circuit being a *master–slave D-type flip-flop*. Because the most common use of the flip-flop is to synchronise logic signals the enable is more commonly called the clock. The internal timing described above is shown in the timing diagram in Figure 2.12.

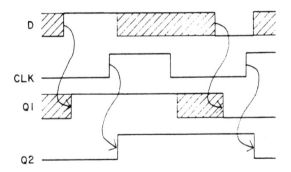

Fig. 2.12 D-type flip-flop internal timing.

2.2.2.4   Summary of flip–flop types

The D-type flip-flop described above is only one of many types, all with different properties and additional features. The designs for some of these are given in the appendix but a summary of their features will be given here in the form of truth tables so that we may refer to them further on in the book. In the truth tables the use of lower case letters refers to the state of the signal in question before the clock or enable signal is applied.

(i) D-latch

D	LE	Q
H	H	H
h	L	H
L	H	L
l	L	L

(ii) D flip-flop

D	Clk	Q
h	^	H
l	^	H

(iii) R–S latch

R	S	Q
L	L	q
H	L	L
L	H	H
H	H	undefined

(iv) R–S flip-flop

R	S	Clk	Q
L	L	^	q
H	L	^	L
L	H	^	H
H	H	^	undefined

(v) J–K flip-flop

J	K	Clk	Q
L	L	^	q
H	L	^	H
L	H	^	L
H	H	^	/q

It is also quite common for an R–S latch to be combined with flip-flops to provide an asynchronous 'setting' and 'resetting' feature.

2.2.3 Timing considerations

2.2.3.1 Combinational functions

We have already noted that there is a delay between data being applied to the inputs of a logic circuit and the resulting output change. This is due to circuit capacitances being charged and discharged, so it follows that identical circuits made on a common process will have the same delay time. In practice process variations make this only approximately true but it is common to quote a figure for the *gate delay* for most logic families. We have already seen how glitches may occur by different logic paths having different delays before being combined in a single output. This effect can sometimes be eliminated by careful design, such as overlapping groupings in a Karnaugh map; nevertheless it is good practice to try to ensure that a minimum discrepancy in number of gate delays occurs between different paths in a logic circuit. That is, that all signal paths contain the same number of gates.

Sometimes it is necessary to ensure that one signal path is longer than another so that signals arrive in the correct sequence. This is often achieved by inserting extra gates or inverters into the 'slower' path. Correct operation may often ensue from this technique but there is always a danger that things can go wrong. Although maximum delays are always specified in logic device data minimum delays frequently are not. Further, the actual delay will also depend on factors such as stray capacitance, supply voltage and temperature, and a positive difference under one set of circumstances can become a negative difference under a changed set. The only safe way to proceed in these circumstances is to use sequential logic.

2.2.3.2 Latches

If we look back to the equation for a D-latch we can see that the two input signals, 'data' and 'enable', both suffer two gate delays between input and output. The circuit diagram, Figure 2.13, shows this more graphically and also shows that LE has two different paths, one inverted and the other non-inverted. If D changes while LE is HIGH then the output will change after a predictable delay. If LE goes LOW while D is unchanging then the output will be latched into a predictable state. Consider what happens though when LE goes LOW while D is also changing; the state of the output is now not easily predicted.

To be sure about the final outcome D must be stable around the time when LE is changing from HIGH to LOW. Part of the specification of a latch must include a definition of the time for which D must not change, but we can estimate what this must be by analysing the circuit. If we start with D and LE both HIGH then AND gates 1 and 2 are both HIGH and gate 3 is LOW. In the latched condition, when LE is LOW and D can be LOW, only gate 3 need be HIGH. Thus to ensure reliable operation of the latch we must be certain that gate 3 goes HIGH before gates 1 and 2 can go LOW. Taking LE LOW will send gate 1 LOW so Q must be HIGH before LE goes LOW. D must therefore be HIGH for at least two gate delays before LE goes LOW. This is called the *setup time*. Because LE is inverted before being gated with Q there is an extra

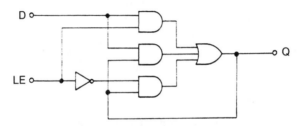

Fig. 2.13 D-latch circuit diagram.

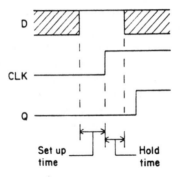

Fig. 2.14 Flip-flop timing definitions.

gate in this path, so D must stay HIGH for one gate delay after LE goes LOW. This is called the *hold time*. These timings are illustrated in Figure 2.14.

The third parameter which must be defined is the length of time for which LE must be HIGH to allow D to be reliably latched. If LE and Q are LOW then all three AND gates are LOW irrespective of D. In order to latch a HIGH, LE must stay HIGH for long enough for a HIGH Q to be fed back before LE goes LOW, otherwise there is a chance that all three gates will go LOW again. This minimum pulse width is clearly equal to two gate delays to allow the feedback to become established.

2.2.3.3 Flip-flops

A simple master–slave flip-flop is merely two latches in series, as we have already established. Reliable operation depends on data being properly latched into the master section of the flip-flop, therefore the same criteria apply to the master as to the simple latch. These same criteria must also be extended to the slave section for the master Q becomes the slave D. Master Q must therefore be established two gate delays before the active clock edge so the setup time becomes four gate delays. The hold time is not affected because the delay through the master ensures that Q will remain stable long after the active clock edge. Because the master and slave use opposite senses of the clock signal for latching, both HIGH and LOW clock states are subject to minimum widths.

The setup time of four gate delays is rather long to make an efficient practical flip-flop so these use R–S latches for the most part. The R–S latch has only a single gate delay between input and output so the setup time for a flip-flop is reduced to two gate delays. A further advantage is that it is relatively simple to build in asynchronous set and reset by adding a parallel latch, so a more versatile, as well as a faster device is obtained.

2.2.3.4 Metastability

Having established rules for the timing of input signals to latches, we can now see what happens if those rules are broken. In Figure 2.15 we have assumed

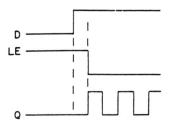

Fig. 2.15 D-latch timing violation.

that the D-input to a D-latch goes HIGH one gate delay before LE goes LOW. The effect of this is to send gate 1 HIGH for one gate delay. Q therefore also goes HIGH but for just one gate delay and this pulse is fed back to gates 2 and 3 sending them HIGH, again for one delay period only. This alternating HIGH and LOW would carry on indefinitely if the gate delays were all exactly equal. In practice they are not, but the closer they are the longer it will take for the oscillation to die away.

In a flip-flop the effect of the master oscillating, when it should be latched, will be passed directly through the slave whose latch is open. Stray capacitance may well cause the oscillation to be smoothed at the device output so the apparent effect will be for the output to lie between the HIGH and LOW states for a relatively long time before a normal level is restored. In extreme cases, where the mark–space ratio of the oscillation matches the difference in gate delays internally, this *metastable* state can continue almost indefinitely.

The lesson is that setup and hold times of flip-flops should not be violated as it can lead to unpredictable results. Unfortunately this is not always possible, for one of the most common uses of flip-flops is to synchronise asynchronous signals. In this case the possibility of metastability should be recognised and the system constructed so that there is an overwhelming probability that the flip-flop will have left its metastable state before the signal needs to be used.

2.3 SEQUENTIAL LOGIC SYSTEMS

2.3.1 System examples

2.3.1.1 Registers

While a latch or flip-flop is capable of storing a single bit of data it is more usual to want to store several bits. For example, in the architecture of Figure 2.9 the address of the next instruction has to be held until the microprocessor is ready for it. A group of flip-flops all controlled by the same clock signal is called a *register*. In this case we have a set of latches rather than flip-flops but the same principle applies. Because each latch has its own input and output it is an example of a parallel register. Parallel registers have some disadvantages, particularly if being made as a single integrated circuit. The most obvious is

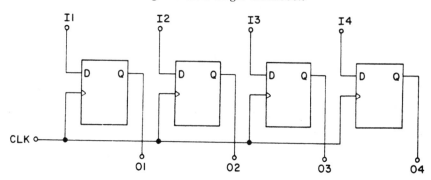

Fig. 2.16 4-bit parallel register block diagram.

the number of connections required, each connection having to be made via a separate pin. For a register of n flip-flops there will be $2n+1$ connections. Furthermore, to connect the register to another device will require one wire for each register bit.

The block diagram of a 4-bit parallel register is shown in Figure 2.16. Suppose that the output of each flip-flop is fed to the input of the flip-flop on its right-hand side, we then have the arrangement of Figure 2.17. On a clock pulse the input to the left-hand flip-flop is loaded, but the previous contents are loaded into the next one, and so on down the line. The data is thus shifted along the register, hence the name *shift register* or *serial register* for this arrangement.

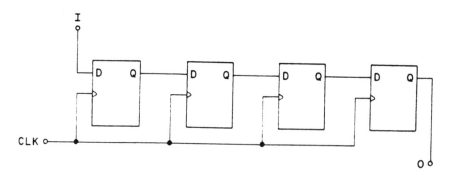

Fig. 2.17 4-bit shift register block diagram.

The number of connections is now reduced to three irrespective of the number of register bits. The chief disadvantage is that n clock pulses are needed to load the register or to read its contents. It is thus much slower to use than a parallel register. Two shift registers can transfer their data along a single connecting wire so hybrid versions may be used to save wiring when two parallel systems need to communicate. In Figure 2.18 the left-hand register is

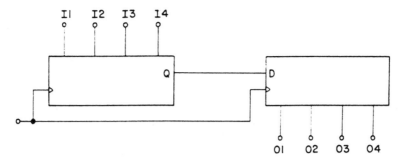

Fig. 2.18 Single wire communication between shift registers.

loaded as a parallel register and the data output as a shift register; the right-hand register is loaded with serial data and read out in parallel.

Another application of shift registers is in arithmetic circuits. If the bits loaded into the register represent a binary number, then shifting them one place results in the number divided by two. Similarly, if the connections are reversed so that the bits are shifted left instead of right, the result of a shift is to multiply by two. These properties are illustrated in Figure 2.19 where the number 13 (binary 1101) is loaded. Note that the bit output in the case of division is the remainder.

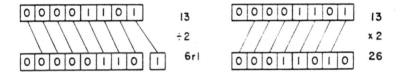

Fig. 2.19 Arithmetic properties of shift registers.

2.3.1.2 Counters

In the summary of flip-flops we noted that a J–K flip-flop inverts the stored bit if both J and K inputs are held HIGH. The same result can be obtained from a D-type flip-flop if the output is inverted and fed back to the input. In either case the output will be a waveform with half the frequency of the clock input. If this output is used as the clock for a second flip-flop the final signal will have a quarter of the frequency of the original clock, and so on. Figure 2.20 shows an arrangement with three flip-flops and the resulting waveforms. In this example the inverted output drives the clock and the outputs follow the sequence 000, 001, 010, 011, 100, 111, 000, etc. using '0' to represent the output LOW state and '1' the output HIGH state between clock pulses. These are just the binary representations of the numerical sequence 0, 1, 2, 3, 4, 5, 6, 7, 0; in other words this circuit, if left to itself, will continuously count from 0 to 7 and

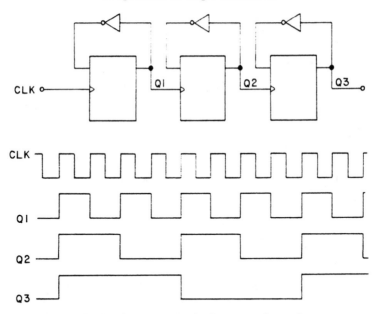

Fig. 2.20 Octal counter circuit diagram and waveforms.

is called an *octal counter*, octal denoting the fact that there are eight counts in the sequence.

By adding flip-flops we can, apparently, extend the counting capacity indefinitely, but there are two constraints. Firstly the counts are tied to powers of two so that only 2, 4, 8, 16, 32, etc. can be used as the counting base; secondly no account has been taken in this analysis of the signal delays through the flip-flops. Every flip-flop which is added to the string will be clocked a little later than the previous one. This restricts the frequency at which the counter can be clocked for the next clock pulse cannot occur until the last flip-flop has had the chance to toggle (change state). This type of counter is called *asynchronous* because all the stages change at different times.

To make a counter in which all the stages change simultaneously, a *synchronous counter*, all the flip-flops must be clocked by the same signal. In this case each stage must tell the following stage when it is to toggle. This may be done with J–K flip-flops by recalling with J and K both LOW the output does not change; in a simple counter each stage toggles only when all the preceding stages are all HIGH, thus by feeding the J and K with the 'AND-ed' outputs from all previous stages it will toggle at the correct time – see Figure 2.21.

The principle of gating outputs together to select particular numbers is used to make counters with bases other than powers of two. To see how this is done we need to investigate a more formal way of representing sequential functions, such as those described in this and the previous section.

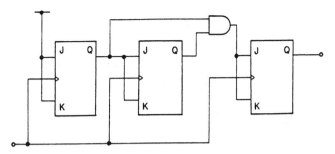

Fig. 2.21 Synchronous counter circuit diagram.

2.3.2 Formal description of sequential systems

2.3.2.1 *State diagrams*

We have used the word 'state' somewhat loosely up to now but will define it more accurately in relation to sequential logic systems. The state of a system is a stable set of logic variables held in a set of flip-flops used to define the system. The set of flip-flops is usually called the *state register* although they may be the same as the *output register* which contains the data intended to be the 'result' to the outside world. A system in which the output is just a function of the state register is called a *Moore machine*; if the output is also a function of the inputs, combined with the state register, the system is a *Mealy machine*.

This sounds long-winded but may be clarified by looking at examples. An octal synchronous counter, as in Figure 2.21, contains three flip-flops which are both state register and output register. The flip-flops define both the eight stable states of the counter (number 0 to 7) and the useful outputs. The function of the counter is to increment the count by one on each clock pulse so the counter must sequence from state '000' to state '001' and then to state '010' and so on. It is possible to list all the states and show the progression from state to state by means of arrows. This is the *state diagram* of the system. Figure 2.22 shows the state diagram for the octal counter with one

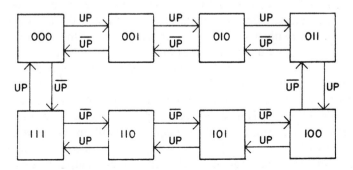

Fig. 2.22 Synchronous counter state diagram.

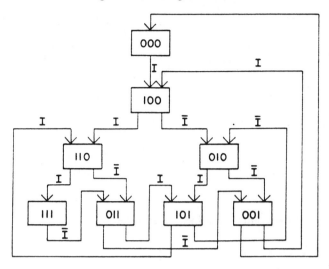

Fig. 2.23 Shift register state diagram.

complication; an input signal 'UP' has been included. When UP is HIGH the counter operates as already described, when UP is LOW it counts in the opposite direction; our counter has become an up/down octal synchronous counter.

It is also possible to draw the state diagram for a shift register. In this case all the flip-flops again form a state register but there is only one output flip-flop, the least significant bit of the state register. The state diagram for a 3-bit shift register is shown in Figure 2.23; unlike the counter diagram it does not describe the circuit function very clearly, nor would it be very easy to design the circuit from the state diagram. What is needed is a more direct way of describing the circuit operation. We will discuss this in the next section.

2.3.2.2   *Karnaugh maps .... again*

The state diagram for a sequential system is analogous to the truth table for a combinational system, so we will see how a sequential system can be described with a Karnaugh map. Because the state which the system takes up depends on the state it is currently in as well as the inputs, each bit of the state register has to be treated in the same way as an input.

Figure 2.24 shows the maps for each of the three flip-flops in the 3-bit shift register; it is clear that Q3 goes HIGH when the input is HIGH, irrespective of any other register bits, Q2 goes HIGH only when Q3 is HIGH and Q1 is HIGH when Q2 is HIGH. It should be noted that the input conditions refer to the conditions before the clock (the *present state*), while the output is the result of the clock (the *next state*).

While combinational logic is capable of only two output levels there are

O2	O1	I O3	0 0	0 1	1 1	1 0
0	0					
0	1					
1	1		H	H	H	H
1	0		H	H	H	H

O3

O2	O1	I O3	0 0	0 1	1 1	1 0
0	0					
0	1		H	H	H	H
1	1		H	H	H	H
1	0					

O2

O2	O1	I O3	0 0	0 1	1 1	1 0
0	0				H	H
0	1				H	H
1	1				H	H
1	0				H	H

O1

Fig. 2.24 Karnaugh maps – 3-bit shift register.

four possibilities with sequential logic. These are HIGH, LOW, TOGGLE and NO CHANGE. The shift register was adequately defined using only HIGH and LOW, but other circuits may be described better using the alternative transitions. For example, the maps for an octal counter using only HIGH and LOW, Figure 2.25, are more complex than the maps using TOGGLE and NO CHANGE, Figure 2.26. This shows that the actual circuit using J–K flip-flops, which allow the TOGGLE and NO CHANGE, will be less complex than the same function built from D-type flip-flops, which allow only HIGH and LOW to be loaded.

UP	0	0	1	1	
O3	0	1	1	0	
O2	O1				
0	0	H	L	H	L
0	1	L	H	H	L
1	1	L	H	L	H
1	0	L	H	H	L

O3

UP	0	0	1	1	
O3	0	1	1	0	
O2	O1				
0	0	H	H	L	L
0	1	L	L	H	H
1	1	H	H	L	L
1	0	L	L	H	H

O2

UP	0	0	1	1	
O3	0	1	1	0	
O2	O1				
0	0	H	H	H	H
0	1	L	L	L	L
1	1	L	L	L	L
1	0	H	H	H	H

O1

Fig. 2.25 Karnaugh maps – octal counter in terms of 'H' and 'L'.

2.3.3   Extension of design example

2.3.3.1   Functional description

Having looked at sequential circuit design we are now in a position to extend the function of the combination lock, which we have been using as a design example throughout the book. As a practical system the lock, as described so far, would be rather inadequate for a real security application. After all by loading only four bits there is a 1 in 16 chance of hitting the correct combination, nor would it take very long to find it by trial and error. By using

O2 \ O1	UP O3	0 0	0 1	1 1	1 0
0	0	T	T		
0	1				
1	1			T	T
1	0				

O3

O2 \ O1	UP O3	0 0	0 1	1 1	1 0
0	0	T	T		
0	1			T	T
1	1			T	T
1	0	T	T		

O2

O2 \ O1	UP O3	0 0	0 1	1 1	1 0
0	0	T	T	T	T
0	1	T	T	T	T
1	1	T	T	T	T
1	0	T	T	T	T

O1

Fig. 2.26 Karnaugh maps – octal counter in terms of 'toggle' and 'hold'.

a sequential system the number of possible combinations can be greatly increased, and incorrect guesses can be detected.

The extended lock will require three digits to be entered in the correct order and only one mistake will be allowed. The system will have four inputs, to allow a BCD number to be entered, plus an 'input' switch to provide a clock signal. The state register will have four bits, two to count the correct sequence, one to detect a mistake and one to signal more than one mistake. The start state has all bits LOW and a correct sequence is detected by the two relevant bits being HIGH.

2.3.3.2 *State diagram*

This exercise will demonstrate two important practical points regarding state
diagrams. The first is that the starting point must be well defined, usually by
ensuring that the state flip-flops take up known levels when power is first
applied. In our example I have assumed that all flip-flops power up LOW. The
flip-flops in the state register are defined as:

Q3 – 'alarm'
Q2 – first mistake
Q1, Q0 – main sequence

The state diagram is shown in Figure 2.27, starting with '0000' at the top left
corner. There are two possible jumps from this state; if the input is '8', the
correct number, then Q0 is set as the first step on the sequence. If the input is
not '8' then Q2 is set to show that a wrong entry has been made. A sequence of
entries '8' – '0' – '9' with no mistakes will cause a progression down the left
hand column until Q1 and Q0 are both HIGH. The same sequence with one
mistake will cause a progression down the centre column again ending up in a
correct conclusion. A second error will cause a jump from the second column
to the bottom right-hand state where Q3, the alarm bit, is set.

The second practical point is that having reached one of the states in the
bottom row there must be a way for the system to return to the starting point,
otherwise the lock could only be used once without turning the power off and
on again. In this case we have provided another input, RESET, which takes
the system back to state '0000' from any of the finish states. From a practical
point of view this is a valid function. An authorised user will need to reset the
lock after use whilst an unauthorised user will probably not want to wait
around if an alarm is set off.

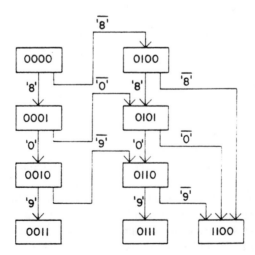

Fig. 2.27 Extended combination lock state diagram.

To summarise, a sequential system must have a well defined entry state and must not include dead ends; every state must have a way in and a way out.

2.3.3.3 Logic equations

In order to construct the system we need to describe the logic in terms which allow us to decide which gates, flip-flops, etc. are needed, and how they should be connected. We can derive logic equations directly from the state diagram by seeing the input and present state conditions required to set each bit **HIGH**. Conventionally the symbol ': =' is used to represent the function 'equals after the clock edge'. Equations for the four state bits are then:

$Q3 \quad := Q2 * /Q1 * /Q0 * /(I3 * /I2 * /I1 * /I0)$
$\quad\quad + Q2 * /Q1 * Q0 * /(/I3 * /I2 * /I1 * /I0)$
$\quad\quad + Q2 * Q1 * /Q0 * /(I3 * /I2 * /I1 * /I0)$

$Q2 \quad := Q2$
$\quad\quad + /Q2 * /Q1 * /Q0 * /(I3 * /I2 * /I1 * /I0)$
$\quad\quad + /Q2 * /Q1 * Q0 * /(/I3 * /I2 * /I1 * /I0)$
$\quad\quad + /Q2 * Q1 * /Q0 * /(I3 * /I2 * /I1 * /I0)$

$Q1 \quad := /Q3 * /Q1 * Q0 * I3 * /I2 * /I1 * /I0$
$\quad\quad + /Q3 * Q1 * /Q0 * I3 * /I2 * /I1 * I0$

$Q0 \quad := /Q3 * /Q1 * /Q0 * I3 * /I2 * /I1 * /I0$
$\quad\quad + /Q3 * /Q2 * /Q1 * Q0 * /(/I3 * /I2 * /I1 * /I0)$
$\quad\quad + /Q3 * Q1 * /Q0 * I3 * /I2 * /I1 * I0$

It is assumed that the reset to '0000' will be achieved by an asynchronous reset to the flip-flops. In that case an equation can be written for the reset, if this signal is called 'R':

$R = RESET * /Q3 * /Q1 * Q0$
$\quad + RESET * Q3 * Q2 * /Q1 * /Q0$

When designing the circuit based on the above equations, it is possible to save hardware by noting that some logic functions, or their complement, are repeated for several of the outputs. For example, the inputs are always decoded to '0', '8' or '9', or their complement, and the state bits Q1 and Q0 are also decoded more than once each in the equations. A possible hardware implementation is shown in Figure 2.28, where an additional output signal, UNLOCK, has been derived as:

$UNLOCK = Q1 * Q0$

This is the signal to show that the correct combination has been successfully entered. It also illustrates the use of an R–S latch to 'debounce' the switch input for the clock signal. Altogether the circuit uses:

22 AND or NAND gates
5 OR gates
8 inverters
4 D-type flip-flops

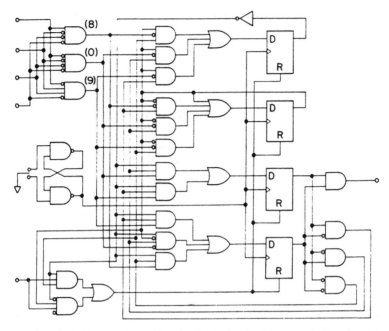

Fig. 2.28 Extended combination lock circuit diagram (discrete logic).

These would require about a dozen integrated circuit packages from a standard logic family to construct the circuit. Later we will examine how this can be reduced to one programmable logic device.

2.4 EXAMPLES

2.1 A combination lock is built to respond to inputs of 4, 5, 7 and F (15). Using, firstly, Boolean algebra, and then, Karnaugh map analysis find the minimum glitch-free solutions for both active-HIGH and active-LOW cases.

2.2 A single bit adder has a carry input and output, as well as 'A', 'B' and 'sum'. Construct a truth table and, with the help of Karnaugh maps, derive a circuit using the least number of AND gates, OR gates and invertors.

2.3 An R–S latch is formed by 'cross coupling' two NAND gates (see Figure 9.16). Show how a latch enable signal could be added to this arrangement, and then develop this into an R–S flip-flop. How can additional feedback change it to a J–K flip-flop?

2.4 By using a state diagram and Karnaugh maps, design a synchronous divide-by-five counter out of D-type flip-flops, and then out of J–K flip-flops. Ensure that the counter will not get stuck if it enters an illegal state.

Chapter 3
Programmable Device Techniques

3.1 MEMORY DEVICES

3.1.1 ROM structure

In Chapter 1 we stated that microprocessors are controlled by a sequence of instructions which determine the internal function at any time. Along with the microprocessor, a method of storing and reading the instruction sequence had to be developed. The obvious way of ordering the instruction sequence is to label the first instruction '1', the second '2' and so on. The instructions can then be stored in some device which will respond to the number '1' with the first instruction; when this has been done the microprocessor can send it number '2' and it will respond with the second instruction and continue in this way until the operation is complete. Conventionally the number of the instruction is called its *address* and it is usually counted in hexadecimal notation, because microprocessors are almost all organised in multiples of 4-bits, or *nibbles*.

The microprocessor architecture referred to in the previous chapter is shown in Figure 3.1; it is called the Harvard architecture. As we noted before there are three groups of connections to the microprocessor, each group being called a *bus*. The *data bus* allows the microprocessor to communicate with the outside world, the *address bus* is the output for the next address and the *instruction bus* is the route back into the microprocessor for the next instruction. We can now look at the internal structure of the 'ROM' which takes the address and sends back the instruction.

ROM stands for *read only memory* which implies that it is fixed and cannot be changed. The ROM must recognise the address which it is receiving and find the instruction corresponding to that address. For simplicity, let us assume that the address bus is a single nibble so that the ROM can recognise an input in the range 0–F. The four inputs must be decoded inside the ROM so that each of the sixteen addresses will give the appropriate output. Referring to Figure 3.2, let us assume that the decoder outputs are HIGH unless the input address corresponding to it is present. For example, an input of '0000' will make the horizontal line labelled '0' LOW but the others (1–15) will be HIGH. The vertical lines will also be normally HIGH but, if any of the bottom row of switches is closed, that line will be taken LOW and cause that output line to be LOW also. Any lines with open switches will remain HIGH.

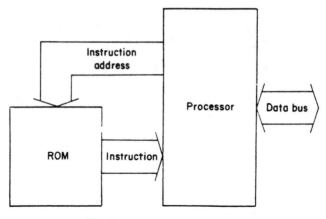

Fig. 3.1 Harvard architecture.

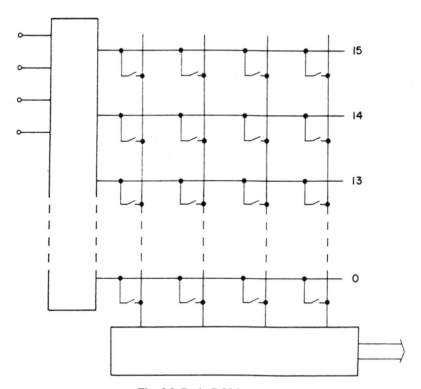

Fig. 3.2 Basic ROM structure.

In practice, simple switches would present problems because the LOW signal could be propagated up the vertical lines to other, non-selected, horizontal lines. We must, therefore, look at the ways in which the switches can be physically incorporated into ROMs in order to prevent this sort of interaction.

3.1.2 Masked ROMs

As we saw in Chapter 1, the simplest electronic switches are either the diode or the MOS transistor, depending on the technology being used. Figure 3.3 shows how either may be used to switch the detected signal; in fact a number of different configurations of the MOS switch have been used by various manufacturers, but the principle remains the same. On the silicon surface contact between the metal interconnections and the active devices is made via contact holes in the silicon dioxide covering the surface. Thus the connection may be permanently made or broken, according to the presence or absence of a contact hole to the appropriate switch. In Figure 3.4 contact is made to the left-hand diode and the output will be HIGH when the address corresponding to that switch is input, while the right-hand one remains open and the output will be LOW.

The information in the ROM is thus stored as the presence or absence of a hole in the silicon dioxide; it being physically programmed into the device during manufacture, according to the information coded into the mask used for opening the contact windows. As we noted earlier, masking processes are only viable for large quantities and take several days, or even weeks, to

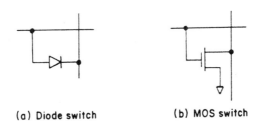

(a) Diode switch   (b) MOS switch

Fig. 3.3 (a) Diode switch; (b) MOS switch.

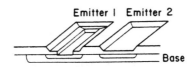

Fig. 3.4 Masked ROM programming.

produce devices. The masked ROM is therefore not suitable for small users or for prototyping, when changes may be needed and tested at short notice. This has led to the development of the PROM or *programmable ROM* which the user can programme himself. The following section describes the various technologies in which PROMs may be fabricated.

3.2 FIELD PROGRAMMABLE ROMs

3.2.1 Metal fuses

3.2.1.1 Fuse structure

In order to make a ROM *field programmable* some way must be found to replace the holes in the contact mask with a switch which can be opened from outside the device. Perhaps the simplest switch is a metal fuse, whose operation in a domestic context is quite familiar. A thin piece of wire is placed in the current path; when the current exceeds a given value the heating effect is sufficient to melt the wire and switch off the current. The switch may only be closed again by replacing the wire, so it is an effective safety device.

On the microelectronic scale electrical connection is achieved by thin films of metal, rather than wire, and a fuse may be fabricated in the same way. As discussed in Chapter 1, aluminium is the usual choice for interconnecting metal; the reasons being that it is easy to evaporate into a reliable thin film, and it has a relatively low resistivity. The requirements for a fuse metal are not quite the same. It must be capable of being evaporated into a thin film, but a high resistivity is preferred as the heat produced by an electric current is proportional to the resistance. Common materials are alloys formed from nickel and chromium, or from tungsten and titanium, while some manufacturers use polycrystalline silicon itself or platinum silicide. A number of different shapes of fuse are used, the commonest being shown in Figure 3.5 as the bar, taper and notch. The advantages of each will be discussed in the section on reliability. A common feature of each is the size, being typically about 5 micrometres wide and 10 micrometres long.

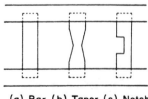

(a) Bar (b) Taper (c) Notch

Fig. 3.5 Fuse shapes: (a) bar; (b) taper; (c) notch.

3.2.1.2 Programming method

Although it is relatively straightforward to fabricate an array of metal fuses and diodes, some way has to be found to blow out the unwanted fuses without damaging any of the other components on the circuit. The typical fusing current is 50–100 mA which could destroy the small signal components within the circuit. Figure 3.6 shows one element in the fuse array with the diode replaced by a transistor; diodes in integrated circuits are formed from transistors in any case. The base of the transistor is driven by the address selection circuit, while the emitter drives a buffer which forms one bit of the output word. The collector is connected to the supply voltage (V_{cc}) so that the drive to the output buffer does not have to be supplied by the address selection circuit. If this transistor is made large enough it could supply the programming current for the fuse, but the output buffer cannot sink this current for its input is usually just the base of a transistor. Besides, there has to be a way of defining that the fuse is being programmed, as opposed to normal circuit operation when it must be made impossible to blow the fuse.

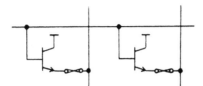

Fig. 3.6 Transistor switch array.

The way round this is to use another property of diodes. Although applying a reverse voltage to a junction increases the potential barrier which charge carriers outside have to climb, it also increases the electric field inside the depletion region itself. Inside the depletion region the silicon behaves as if it were intrinsic, or undoped. In this situation thermal agitation causes a few carriers to be produced spontaneously. The electric field sweeps them out of the depletion region and they appear as *leakage current* to the outside world. At a critical value of electric field they acquire enough energy to create further carriers in collisions during their passage through the depletion region. This multiplication causes a rapid increase of leakage current and is termed *avalanche breakdown* of the junction. It occurs at a voltage which depends on the doping levels on either side of the junction, and is called the *breakdown voltage*.

A rather different mechanism can occur in very heavily doped junctions; this is called *zener breakdown*. In this case the mechanism is somewhat different and results from heavily doped junctions having a very narrow depletion region. Carriers with sufficient energy can *tunnel* through the depletion region; this is a process which can be described mathematically by quantum mechanics. Physically, what happens is that the electric field breaks the chemical bonds holding the atoms together and allows current to flow through the depletion region.

The emitter–base junction can be used to form a well defined *zener diode* and is used in bipolar PROMs to enable the fuse blowing circuitry. Fusing specifications call for voltages above the normal 5 V to be applied to V_{cc} and output pins. The V_{cc} overvoltage may select larger transistors to supply the fusing current; although the transistors used for normal operation could be made large enough this would cause a speed penalty, so it is usual to have a second set of drivers for fusing only. The output overvoltage will cause the output buffers to be by-passed and enable a set of transistors which can sink the fusing current. The fundamental circuit principle is shown in Figure 3.7; actual circuits used may differ in detail as various techniques have been devised to improve programming performance.

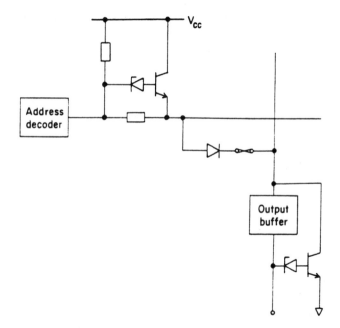

Fig. 3.7 Basic programming circuit.

3.2.2 Diode fuses

3.2.2.1 Fuse structure

Although metal fuses are well established, indeed they have been used in production devices since 1970, they do increase substantially the size of the device compared with an equivalent mask programmable circuit. The fuse itself is an extra component on the silicon surface associated with each diode in the array, while the mask programming is built into the diode structure itself. The diode fuse overcomes this drawback by making the programmable element part of the diode itself.

Each cell consists of a diffused transistor with the base connection left open-circuit; see Figure 3.8. As described in the previous section, a reverse voltage in excess of the breakdown voltage will cause a large current to flow through the junction. This property is used in zener diodes when the current flow is restricted to limit power dissipation in the junction, thereby avoiding any damage to it. In a diode fuse sufficient current is allowed to flow for the junction to melt locally; aluminium from the surface is carried by the current flow into the junction, where it alloys with the silicon. The alloy thus formed has a lower resistance than the silicon junction, so the power dissipation decreases and damage to the junction is limited to the production of a short circuit.

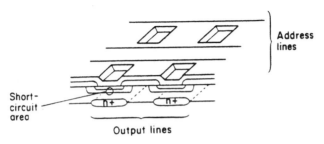

Fig. 3.8 Diode fuse.

The process is known as *avalanche induced migration*, or AIM; there is an enhanced version called *diffused eutectic aluminium process*, or DEAP, in which additional silicon and metal layers are deposited over the diode in order to control the alloy formation more closely.

3.2.2.2 *Programming method*

As with metal fuses, the fusing circuitry is enabled by overvoltages feeding through zener diodes. Unlike metal fuses, current flows after the fuse has been blown. The time taken for either metal fuses or diode fuses to blow is not exactly definable. It is normal for the programming time for a metal fuse to be set much longer than the actual time taken for it to fuse. No current flows after the fuse has blown, so no additional stress is suffered by the fusing circuitry. If this procedure was adopted for diode fuses, the components in the fusing path would be carrying the fusing current for much longer than necessary. Thus the programming method calls for a number of short pulses, with a verification after each pulse to see if it is possible to end the sequence. Once a short circuit has been detected it is usual to apply a final pulse to burn it in.

With so much heat being dissipated and so much current flowing it might be expected that the collector–base junction would be damaged also. The collector–base junction has a larger area than the emitter–base so the heat generated there will not cause such a large temperature rise; furthermore, the voltage drop across the collector–base is an order of magnitude lower because it is forward biased. Nevertheless, one of the most important considerations in designing the programming specification is to ensure that the remaining diode is not degraded when the diode fuse is short-circuited.

3.2.3 MOS floating gate cell

3.2.3.1 *Floating gate structure*

Both the metal fuse and the diode fuse share the need for a large current pulse to blow the fuse, and hence programme the PROM. As we noted in Chapter 1, only the bipolar process lends itself readily to the passage of high currents. Fuse link and AIM PROMs are almost exclusively made by that process; the one exception is where silicon itself is used as the fuse material. Some CMOS PROMs are made with silicon fuses because a higher resistivity is achievable than with metal, and therefore lower fusing currents can be used, but the resistivity is harder to control so this technology has been less successful.

A completely different memory cell has been developed for MOS devices based on the operation of the MOS transistor itself. The structure, called *floating gate*, is illustrated in Figure 3.9. There are two gates, the upper one is the normal gate connected to the output from the address decoding circuit. The lower gate, on the other hand, is isolated electrically from it and from the channel. The action of this floating gate is to modify the threshold voltage of the MOS transistor, depending on whether or not it is charged. Charging takes place by injection of high energy electrons onto the gate through the

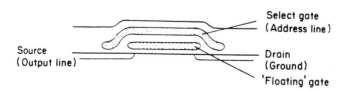

Fig. 3.9 MOS floating gate cell.

surrounding oxide; the oxide prevents the charge from escaping during normal operation. Figure 3.10 shows the effect of the two threshold voltages on conduction through the cell. When the floating gate is charged it raises the threshold to a level such that the transistor cannot conduct when 'normal' voltages are present on its gate. Thus the MOS transistor acts as a switch which can normally be turned on and off by the address selection circuit, but is permanently off after programming.

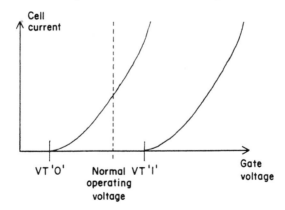

Fig. 3.10 MOS cell thresholds.

3.2.3.2 *Programming methods*

As stated above, the floating gate is charged by injecting high energy electrons through the surrounding oxide. The oxide layer between the two gates acts as a potential barrier to electrons, but may be overcome by applying a high voltage (26 V originally) between the control gate and the silicon substrate. Although there is no direct connection to the floating gate, it will acquire some intermediate potential which remains when the voltage source is removed. The cell to be programmed is selected by addressing it through the address decoder and the output pins. Some MOS PROMs incorporate a separate pin which enables the addressing circuits in the programming mode.

The method of programming has developed since the introduction of MOS PROMs. The first devices called for the charge to be built up gradually by cycling through the addresses with a short (< 1 ms) pulse at each address. This minimised the problems caused by heating and capacitive coupling between adjacent cells. Improvements in design and processing allowed this early method to be replaced by what is now the standard method; that is to use a single 50 ms pulse on each cell. This leads to an inconveniently long time to programme a single PROM; nearly 7 minutes for a PROM with 8192 addresses. To improve matters so called 'intelligent' algorithms have been developed.

Intelligent methods rely on using a series of short pulses which are counted by the programming equipment until the cell is detected as being programmed. Usually a 6 V supply is used to increase the threshold voltage which must be overcome. Once the cell has become charged enough to register as being programmed, a final programming pulse is applied to ensure that the floating gate is fully charged. The length of the final pulse is calculated from the number of short pulses first applied. This is usually two to three times quicker than the standard 50 ms pulse, which has to be set long enough to ensure reliable programming for all cells in the PROM. In practice, if several PROMs are being given the same programme a *gang programmer*, which can accommodate 8 or 16 PROMs, is used.

3.2.3.3 Erasing MOS PROMs

Once a metal fuse or diode fuse has been blown it is impossible to reverse it to make the connection again. The floating gate is a different case, however, since no irreversible process is involved in charging the cell. At first sight the obvious way to reverse the process is to apply a negative voltage across the floating gate, in order to discharge it. The result of doing this might be to finish up with the gate charged negatively, which would leave the threshold voltage negative; this would allow current to flow irrespective of the positive voltage applied to the control gate. In any case, it would be impossible to determine exactly the end point of the discharging process to arrive at a satisfactory unprogrammed state.

A way must be found to allow the stored charge to be removed completely without introducing any spurious charges or damaging the cell. If electromagnetic radiation falls on a semiconductor the electrons will absorb energy, provided that the energy of each quantum is enough to allow the electron to free itself from the bound state in which it is held. This principle is used in photodiodes where energy absorbed by a p–n junction produces electron–hole pairs which, in turn, create a large increase in the junction leakage current. Infra-red quanta are sufficiently energetic to excite a photodiode, but the electrons in the floating gate have a much higher potential barrier to cross. The MOS cell needs ultraviolet radiation to supply sufficient energy to the stored charge for it to cross the surrounding oxide.

MOS PROMs supplied in a package with a window to allow the data to be erased by ultraviolet radiation are called *erasable PROMs* or *EPROMs*. Care has to be taken to give EPROMs sufficient time under the ultraviolet otherwise the floating gates will not be completely discharged; then the resulting threshold will be indeterminate and cause incorrect data to be sensed. EPROMs have a great advantage over fuse-link PROMs when developing programmes for microprocessors. It is very common to want to make changes to programmes, or to try small parts of a programme, in which case the ability to reuse EPROMs makes programme development much less costly than would otherwise be the case.

3.2.4 Electrically erasable PROMs

3.2.4.1 Structure

Figure 3.11 shows that the electrically erasable cell is very similar to the floating gate cell. Instead of the floating gate, it uses charge stored in surface state sites at the interface between a layer of silicon nitride and silicon dioxide. Wherever there is an interface between different materials the disruption

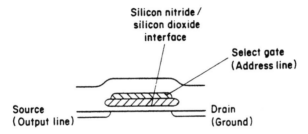

Fig. 3.11 MNOS erasable cell.

caused to the crystal lattice on an atomic scale leaves sites which can be filled by free charges passing through the interface. As with floating gates, there is a large potential barrier for charges to overcome if they are to travel from the control gate to the interface sites. This cell is commonly called an *MNOS cell* to underline its metal nitride oxide silicon structure.

3.2.4.2 *Programming and erasing method*

MNOS cells are programmed in much the same way as floating gate MOS cells, that is by applying a short voltage pulse between the control gate and substrate of the cell being charged. Electrons are attracted into the empty interface sites and raise the threshold voltage of the transistor. Because it is surrounded by silicon nitride and silicon dioxide, which are both good insulators, the charge stays in place until it is forced to move.

The difference from the floating gate cell comes in the erasing method. There are only a limited number of interface sites and they can only accept charges of one polarity, so it now becomes feasible to remove the charge by applying a reverse voltage to the control gate. Once the sites have been emptied no more charge can move and the cell is in the unprogrammed state. Apart from being, arguably, more convenient to erase the device electrically rather than by ultraviolet irradiation, it means that the cells can be erased an address at a time so that complete erasing and reprogramming is not necessary whenever a change to the stored data is needed.

3.2.5 Antifuse connections

So far we have described connections in PLDs being made by metal fuses, short-circuit diodes and MOS transistors. These all use a significant area on the silicon surface, either because of their lateral dimensions or because they use successive diffusions to define their geometry. A fourth technique has been designed, with much smaller dimensions, making it suitable for use in LSI devices where there could be over 100 000 potential connections. This 'antifuse' is called a *PLICE (programmable low impedance circuit element)*.

The antifuse structure, shown in Figure 3.12 in cross-section, is open circuit until it is blown. Connection to the bottom of this vertical structure is via

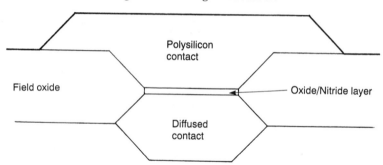

Fig. 3.12 'Anti-fuse' structure.

highly doped n-type diffusion, which has a thin dielectric layer grown over it. The dielectric is a sandwich of silicon nitride between silicon dioxide layers, and it is covered by a layer of polycrystalline deposited silicon, which forms the upper contact layer. A short pulse (<1 ms) of 18 V will rupture the dielectric and give a resistance of less than one kilohm across the antifuse.

It is claimed that reliable fusing is obtained with pulse currents below 10 mA. As with the diode fuse, the danger is that after fusing a short circuit is produced, so steps must be taken to limit the amount of energy dissipated in the fusing process to avoid damaging other areas of the circuit.

3.3 PRACTICAL CONSIDERATIONS

3.3.1 Bipolar PROM designs

3.3.1.1 The diode array

Commercially available bipolar PROMs range in size from 32 × 8 (5 inputs, 8 outputs) and 256 × 4 (8 inputs, 4 outputs) to 8192 × 8 (13 inputs, 8 outputs). Design of the diode array affects both the performance of the PROM and, because of the effect on chip size, the ultimate cost of the device. Clearly it must be possible also to supply enough current to blow the fuses in the finished structure. Moreover the problems involved in designing a small PROM are likely to be a lot less than those for a large PROM. Figure 3.13 shows the actual structure of a 32 × 8 PROM.

Depending on the pattern of HIGH and LOW voltages on the five input lines, one of the horizontal lines, or *rows*, will be set to a low voltage. Those crossovers which still have an intact fuse will take the corresponding vertical line, or *column*, to a low voltage through the diode at the crossover. All the other diodes on that column are reverse biased, so none of the other rows will be affected by the columns which have been pulled LOW. The pattern of HIGHs and LOWs is taken to the outputs via the output buffers, which provide enough current drive to interface with the outside world.

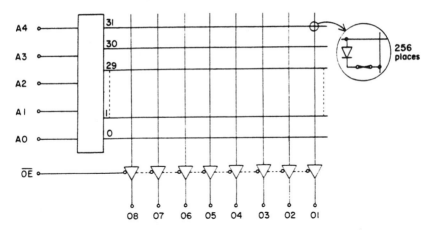

Fig. 3.13 32 × 8 PROM structure.

This pattern may not be extended indefinitely in order to make a much larger PROM. For example, an 8192 × 8 PROM would require 8192 rows and 8 columns which would make for a very difficult connection problem on the surface of the PROM chip. It would also cause the area of the chip, and its power consumption, to be very large because each row signal has to be derived from the address lines by some kind of logic circuit. The way round this has been to change the shape of the PROM diode matrix to 256 rows by 256 columns. Thus only eight of the input lines are used to select the rows, the other five inputs select one group of eight columns from the 256. This structure is illustrated in Figure 3.14.

There is another advantage to this structure. As mentioned earlier, there is a small leakage current associated with a reverse-biased diode; if there are 8192 rows then it is possible for a column, when LOW, to have to sink the leakage current from 8191 diodes. A similar problem would exist if all the input selection were applied to the columns. Thus there are good electrical reasons, as well as geometric ones, for making the diode array as 'square' as possible.

3.3.1.2 Speed considerations

One of the chief advantages of bipolar PROMs is that they are relatively fast; that is it takes only a few tens of nanoseconds for the outputs to change after changing the inputs. Any aspect of the design which adversely affects the speed therefore needs careful consideration.

There are three sections of the PROM which have an effect on the delay of the signal. The first delay is from the inputs to the row lines; the second is the time taken for the column lines to change; lastly there is delay from the columns to the outputs. Both the first and last sections are dependent on the fabrication technology in much the same way as any other logic chip. The main difference is between small PROMs which use only input row selection

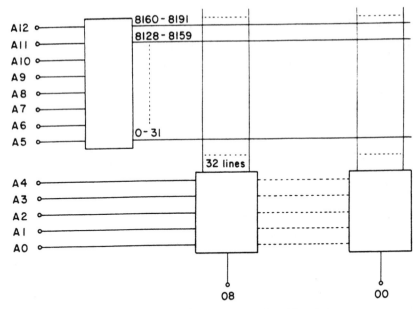

Fig. 3.14 8192 × 8 PROM fuse matrix addressing.

and larger PROMs which have extensive logic on both rows and columns. Thus a large PROM will be slower than a small PROM simply because of the fact that it contains more complex logic circuitry.

More important, in a large PROM, is the delay through the diode matrix itself. This is simply a matter of charging and discharging the capacitance associated with the diodes. Clearly, the larger the PROM, the more diodes it uses and so the matrix capacitance will be higher. In small PROMs the matrix delay is insignificant compared with the other delays, but for larger PROMs the matrix delay starts to predominate. The choice of diode type is therefore important for large PROMs. The type of diode with the smallest capacitance is one formed by a metal–silicon junction, the *Schottky diode*. This has a lower capacitance than any of the various diffused diodes and is therefore commonly used in large PROMS. Its main disadvantage is that it suffers a large voltage drop when high currents are passing, as in the programming phase. Careful geometric design of PROMs is therefore necessary and two layers of aluminium interconnection are essential to reduce other series resistance to a minimum.

3.3.1.3 Power consumption

While device speed is a very strong feature of bipolar PROMs, their power requirements cannot be ignored, the consumption being limited by the temperature rise allowable after packaging. Power is consumed in both the logic circuits and the diode matrix and the speed of the device is broadly determined by this power. After all, if the current supply to the columns is

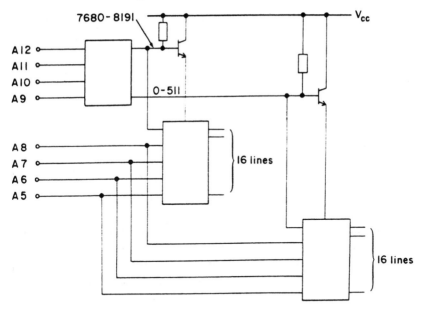

Fig. 3.15 Internal power-down in PROMs.

reduced the capacitance will be charged and discharged at a slower rate, and similarly with the logic circuitry.

The power can be limited by supplying current to only that part of the PROM which is actually being used. In the same way that a certain combination of input signals selects one row in the diode matrix, the input signals can be used to select which section of the PROM is being powered-up. Naturally this affects the speed to a certain extent; the signals cannot pass through unpowered sections of PROM and must wait until they are supplied, but this has less effect than reducing the current supplying each section. Figure 3.15 shows how this scheme works in practice.

3.3.2 MOS PROM designs

3.3.2.1 The cell array

MOS PROMs have an even more compelling reason than bipolar PROMs for using a square memory cell array. Mention was made previously of the capacitance of the diodes, and how this became more significant with larger array sizes. The memory cell in an MOS PROM is just a small capacitor and so the array design needs particular care if performance is not to be impaired too greatly. Using a square array is one step towards minimising the capacitative loading in the memory cell array.

The usual way to reduce capacitance in integrated circuits is to make the components smaller. This has the side effect of making the signals from the memory cells smaller, so techniques more usual in RAMs (random access memories) may be used to read the memory cells. In particular it is possible to incorporate unprogrammed dummy cells alongside the memory array and use these as a reference voltage for a sense amplifier. The amplifier output provides the data output for the PROM.

3.3.2.2 *Performance considerations*

We have previously seen that speed and power are closely tied together for integrated circuit performance. Some of the techniques used in bipolar PROMs are also applicable to MOS devices. Again only the section of PROM currently addressed needs powering so large PROMs can work at the same current levels as smaller PROMs without a significant increase in overall power. In order to speed up the response to address changes, internal circuitry can detect a change in the inputs and power up all the address detectors before the new address has become established. This same signal can precharge the sensing circuits to speed up changes in the output lines.

The main advantage that can be obtained from MOS is the use of CMOS in the peripheral logic circuits. As seen earlier, CMOS circuits consume very little power, but are still able to operate at relatively high speeds when driving other internal components in an integrated circuit. In this way power consumption may be reduced considerably without sacrificing speed. The only adverse effect is to make the chip area larger, which increases the cost of the finished device. The larger the PROM the smaller is the proportion of area taken by the peripheral logic, so CMOS tends to be used more for the larger MOS PROMs where the cost penalty is less.

3.3.3 Quality and reliability

3.3.3.1 *Programmability*

One of the parameters which cannot be tested in a fused PROM is whether or not the fuses can be programmed. The only way to test a fuse is to blow it after which it cannot, of course, be used again. The upshot is that a proportion of the PROMs will not programme correctly. There are some measures which manufacturers can take to minimise this programming loss.

As far as the manufacturing itself is concerned then process control is the chief tool to better quality. The thickness of the fuse metal and dimensions of the fuse can be measured to ensure that the fuse can be blown by the specified current. The diffusion parameters will also ensure that the breakdown voltage is within the correct limits for the zener diodes, and for the fusing of diode fuses. Assuming that these are kept under control then correct application of the programming voltages should result in perfect programming.

In practice this will not happen because of the random manufacturing defects which can occur. These can prevent transistors from operating correctly and, for example, cause the address selection logic to operate incorrectly. Faults can also affect the metallisation causing short circuits or open circuits; these may cause the wrong fuse to be blown or a fuse not to be blown at all. These faults can be minimised by clean manufacturing conditions and by quality assurance inspections during the manufacturing process.

It should be noted that all integrated circuits are affected by these faults but it is usually possible to eliminate the rejects by electrical testing. Most PROM manufacturers now incorporate a test row and column of fuses which enable the fusibility to be tested and, to a certain extent, the operation of the addressing circuits. There is also a useful spin-off from this. In an unblown PROM the outputs will remain unchanging irrespective of the inputs, so it is possible to test only one of the output levels, and impossible to test the delay time through the PROM. If the test fuses are blown then these measurements can be made by addressing the extra memory locations.

Because the memory cells of MOS PROMs can be erased and reused, it is possible to programme the device with a test pattern and then erase it before it is delivered to a customer. Naturally this will allow all the d.c. and a.c. parameters to be fully tested also. This testability means that it is not necessary to provide an extra row and column, however there is a way in which extra memory cells can be used.

As memory arrays are made larger the chance of a manufacturing defect causing a faulty cell becomes significant. By incorporating test circuits and a register of faulty cells it becomes possible to substitute spare cells for the faulty ones during operation of the PROM. The slight increase in area is more than compensated by the increase in the number of good PROMs which is obtained by this technique.

3.3.3.2 Data retention

Reliability of a PROM may be assessed as the length of time for which data is retained in the memory cells. All integrated circuits are subject to failure mechanisms which may stop them operating correctly at some time. Part of the process of improving integrated circuits is the reduction of the effect of failure mechanisms thereby increasing the useful life of these devices. PROMs have additional potential failure mechanisms which must be taken care of if they are to have a useful life comparable with other integrated circuits.

Considering first bipolar PROMs, it should be noted that during normal operation there will be some current passing through the fuses. All the columns are driven by current generators and, referring back to Figure 3.6, it can be seen that this current will pass through the fuse and diode into the selected row line. If the current is significant compared with the fusing current then there will be a risk that it will blow the fuse during normal operation. Analysis of the fusing mechanism shows that there is a critical current density above which rapid melting occurs, but below which the metal melts only slowly. This well-known relationship is shown in Figure 3.16 and PROMs are designed so that the current density during operation is well below the critical level. All the major bipolar PROM manufacturers have published life test results which show that the reliability of bipolar PROMs is no different from

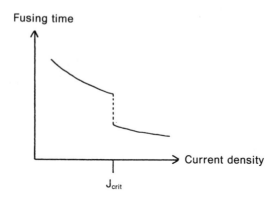

Fig. 3.16 Fusing time vs. current density in nichrome.

other integrated circuit families, and that none of the failures is due to a fuse failure.

It might be thought that the shape of the fuse would have a bearing on both fusibility and reliability. A bar shaped fuse is probably easier to make because alignment is less critical during the manufacturing phase. To be effective, a notched or tapered fuse needs the narrow section to be well removed from the contact area so that fusing occurs away from the aluminium tracks. A bar fuse will always melt in the middle because heat conduction will be symmetrical and ensure that this is the hottest part. There is probably no advantage in shaping the fuse to define where the hottest part will be. Experience has shown that the process quality and electrical design will determine which PROMs are the more reliable.

Data integrity of MOS PROMs is dependent on different factors. As we have seen, programming is by application of a relatively high voltage to the memory cells. During normal operation the devices will not see such a high voltage so there is little risk of losing data. MOS PROMs are designed to be erased, of course, so therein lies the greatest risk of corrupting the stored data. To erase the data fully, normally requires a prolonged dose of ultraviolet radiation but even a short dose will cause the threshold voltage of the memory cells to change. Erasure is a cumulative effect; that is a continuous exposure is not required but the total accumulated time will cause the same effect.

In order to protect an MOS PROM from spurious erasure it is usual to cover the window in the package with an opaque lid or label after programming. This is because sunlight and fluorescent lights contain small amounts of ultraviolent radiation and could cause data to be lost. More important, particularly in military applications, is the effect of ionising radiations such as X-rays and even natural radioactivity. Electrons in the floating gate which are struck by a photon or particle will absorb enough energy to jump the insulating gap and start the erasing process. Even the recently introduced 'one-time programmable' MOS PROMs, which are simply EPROMs supplied in a standard opaque package, are not fully protected from these effects.

3.3.3.3 *Metal fuse regrowth*

A particular aspect of data retention which has caused concern to prospective PROM users in the past has been the ability of the blown fuses to heal themselves. Some very early PROMs did suffer from the problem of fuse regrowth, when a fuse which had been blown was subsequently found to be intact. It has been known for some time that metal in thin films can move under the influence of an electric field. This is one of the commonest failure mechanisms in integrated circuits and is known as electromigration.

Referring back to Figure 3.16 it can be seen that there are two regions where fuses can be blown. High current densities cause a rapid melting of the metal and surface tension pulls the metal away from the hot spot to leave a very clean gap in the fuse. Low current densities cause very local melting resulting in a ragged gap, often with islands in the middle, with small separations which can be filled again when a low voltage is applied. It is clearly important that the programming method ensures that the current density is high enough to give proper fusing. It is up to the manufacturer to design the PROM and specify the programming method, but the user must also be careful that he follows recommended procedures.

The two most important safeguards are to ensure that good contact is made to the PROM and that a PROM which does not programme first time is discarded. If poor electrical contact is made to the PROM then the resulting voltage drop may be large enough to prevent the critical current density from being achieved in the fuses. Similarly, failure to programme first time may be due to a faulty current generator in the programming circuit. Repeated programming attempts may cause the fuse to be blown in its unreliable mode, making it liable to regrowth during subsequent use. By following the programming specification implicitly, perfectly reliable fusing should take place.

3.4 EXAMPLES

3.1 What factors should influence the choice of PROM in the following applications:
(a) A large program memory for a microprocessor operating at 10 MHz?
(b) A look-up table in a portable measuring instrument?
(c) A code converter in a high-speed mainframe computer?

3.2 What steps should be taken to ensure data integrity in both bipolar and MOS Proms?

Chapter 4
Simple PLDs

4.1 PROMs

4.1.1 Programmable logic concept

We have seen, in previous chapters, how electronic components can be built into logic components and how these logic components can be built into useful logic systems. In all this discussion it was assumed that the logic components were fixed and unalterable. That is, if we wanted a NAND gate we could buy a small integrated circuit which would have one or more NAND gates inside it, and that we could interconnect several of these packages by copper tracks on a printed wiring board to make the desired logic function. By anticipating the requirements of electronics engineers the integrated circuit manufacturers have also produced more complex functions, by transferring the copper wiring to aluminium tracks on the silicon surface. The designer is still constrained to using those functions which the manufacturer has provided for him.

The only way around this has been in those cases when an engineer has wanted large quantities of the same function, in which case it has been possible for the manufacturer to make him his own circuit by using custom or semi-custom techniques. We have seen one case, however, where the engineer has been able to modify the integrated circuit to meet his own particular requirements. That is where he is using microprocessors and the operating program can be held in a PROM. In examining the structure of a PROM we found that it was built from logic circuits which are used to address a programmable memory matrix. There is no reason then why the PROM itself cannot contain a logic function.

This is the concept of programmable logic. For the remainder of the book we shall develop this concept into practical devices which have the generic name *programmable logic devices*, shortened to PLDs. We shall start by looking at PROMs themselves.

4.1.2 PROMs as logic devices

As we have seen earlier, a convenient way of representing a logic function is by means of a truth table. By way of example we used, in Section 2.1.1, a combination lock circuit with two valid combinations of four input bits. We have also noted the convention whereby a HIGH logic level represents the binary digit '1' and a LOW level '0'. Thus the truth table could be written as:

Input	Output
9	1
A	1

'A' is the hexadecimal notation for decimal number '10'. Rather than build this circuit out of discrete logic gates it would be just as valid to make it from a PROM. The PROM would be programmed so that addresses '9' and 'A' gave an output of '1' while all other addresses gave an output of '0'.

When looked at in basic terms, there is no difference between a combinational logic device and a memory. Both are designed to give a fixed output in response to a given input combination. The only difference is in their application. A memory has a rather narrow application, particularly, but not exclusively, in storing microprocessor operating programs. Logic devices, of course, have a much wider application.

More relevant than the truth table is the Karnaugh map. Each cell in a Karnaugh map represents a unique combination of input logic levels; indeed every possible combination is represented by one of the cells. In a PROM every possible combination of input signals will drive one, and only one, row of the memory matrix. Thus there is a one-to-one relationship between map cells and memory cells and it is possible to derive the memory table of a PROM directly from the Karnaugh map of the logic function. In practice this would be tedious apart from the simplest of functions involving small PROMs, so this procedure is usually carried out by a computer program. Details of software capabilities and availability are discussed in a later chapter.

4.1.3 Practical limitations of PROMs

The above discussion begs the question as to why all logic circuits have not been replaced by PROMs. Consider the advantages; all the logic functions you need can be made from one, or at least a small number of PROMs, so buying and stocking are much simpler while the designer is not restricted to set logic functions in a particular integrated circuit. Some reduction in the number of packages is bound to occur as logic systems do not need to be designed with individual gates, but by the relationship between inputs and outputs. If a mistake is made in the design it can probably be corrected by making a small change to the contents of the PROM instead of changing the printed circuit board.

These are all valid points, but there is a price to pay. The complexity of PROMs makes them more costly than simple logic integrated circuits; a 256×8 PROM contains over 100 gates plus a 2048 fuse matrix, so on purely economic grounds a considerable reduction in package count is necessary. The performance of a PROM is also inferior to discrete logic circuits in the same technology. For example a typical PROM causes about 30 ns of signal delay for a current consumption of over 100 mA (assuming bipolar technology). A discrete logic circuit with the same number of gate stages would cause about 10 ns delay with less than 50 mA. Again, only by condensing the logic into fewer packages does it become viable to design with PROMs. There are areas where PROMs come into their own as we shall see later.

4.1.4 PROM availability

Part of the function of this book is to be a practical guide to the designer of logic circuits so there will be lists showing what is currently available and giving some indication of performance. PROMs can, as we have stated, be used as memories or as logic device. This is an artificial distinction based on the use to which the PROM is being put, rather than an intrinsic property of the PROM. Nevertheless, a novel method of naming PROMs was devised by Monolithic Memories Inc., now a subsidiary of Advanced Micro Devices Inc., to emphasise their logic capability. They coined the term *programmable logic element* or *PLE*, and describe each PLE as an *mPn*, where '*m*' is the number of inputs and '*n*' the number of outputs. The '*P*' refers to the output level defined in the Karnaugh map. Some other PLD families are restricted to active HIGH outputs or active LOW outputs; clearly a PLE can have its outputs defined as active HIGH or LOW and is thus classified as having 'programmable' polarity outputs.

Tables 1 and 2 show the PLEs available at the time of writing. Bipolar PLEs are still chiefly used for the smaller designs although there is some overlap with CMOS at the top end; this enables the lower power of CMOS to be used where this is desirable. The larger CMOS devices are targeted more at true memory applications than logic, as is shown by the jump in delay time after the 13*P*8.

Process	PLE	Memory	I/Ps	O/Ps	Delay	Current
Bipolar	5P8	32×8	5	8	15 ns	100 mA
Bipolar	8P4	256×4	8	4	25 ns	120 mA
Bipolar	8P8	256×8	8	8	30 ns	140 mA
Bipolar	9P4	512×4	9	4	30 ns	130 mA
Bipolar	9P8	512×8	9	8	30 ns	155 mA
Bipolar	10P4	1024×4	10	4	35 ns	140 mA
Bipolar	10P8	1024×8	10	8	35 ns	170 mA
Bipolar	11P4	2048×4	11	4	35 ns	150 mA
Bipolar	11P8	2048×8	11	8	35 ns	175 mA
Bipolar	12P4	4096×4	12	4	30 ns	155 mA
Bipolar	12P8	4096×8	12	8	35 ns	175 mA
Bipolar	13P8	8192×8	13	8	40 ns	175 mA
ECL	8P4	256×4	8	4	17 ns	160 mA

Table 1   Bipolar PLEs.

Part No.	PLE	Memory	I/Ps	O/Ps	Delay	Current
7C281	10P8	1024×8	10	8	30 ns	90 mA
27C291	11P8	2048×8	11	8	35 ns	75 mA
27CX321	12P8	4096×8	12	8	35 ns	40 mA
27HC641	13P8	8192×8	13	8	35 ns	75 mA
27C128	14P8	16384×8	14	8	120 ns	30 mA
27C256	15P8	32768×8	15	8	120 ns	30 mA
27C512	16P8	65536×8	16	8	120 ns	40 mA
27C010	17P8	131072×8	17	8	120 ns	40 mA
27C1024	16P16	65536×16	16	16	120 ns	160 mA

Table 2   CMOS PLEs.

There are useful applications for even the largest EPROMs as logic devices, however.

The major activity in new designs is in CMOS, because the CMOS memory cells are much more compact than bipolar, and consume less power. Also there is the problem of current distribution through the aluminium tracks. One mm of a typical aluminium track has a resistance of 1.5 ohms; there could well be problems in supplying the 50–100 mA fusing current to the far corners of a large bipolar memory chip, without suffering unacceptable voltage drops.

Although there are applications which can be handled more easily by PLEs, the majority of logic circuits are handled better by other means, as we shall soon see. First it is worth looking at some applications where PROMs have been used to advantage over other methods of logic implementation.

4.1.5   PROM applications

4.1.5.1   Address decoding

We have already seen how some microprocessor systems use a multiplexed data and address bus to save pins. The address output is used to find the next instruction in the program, but it is also used to define the location where temporary storage of data can be made. *Random access memory* or RAM is used for temporary data storage. A detailed description of RAM circuits is beyond the scope of this book; suffice it to say that as well as being readable in the same way as a ROM, data may be written into each memory location. There is usually an input pin to tell the RAM whether it is read mode or write mode. RAMs and ROMs also have an input called 'output enable'. When this input is inactive the output exhibits a high impedance which is neither HIGH nor LOW. This third output state, sometimes called *tri-state*, enables the outputs of several devices to be connected together without affecting each other, provided that only one is enabled at any one time. In this way large blocks of ROM or RAM may be built up from smaller units.

In a microprocessor system each small unit is given an address, or range of addresses, for enabling by the processor; thus the address has to be decoded to find which unit is being selected. The decoding may be done with discrete logic devices in a similar manner to the address decoder inside a PROM, but these may often be replaced by the PROM itself. In the system shown in Figure 4.1 the program needs three 8192×8 PROMs to store it, there are 16 384 bytes of RAM, and data interfaces to the system, called I/O, are also given addresses so that they can be selected in the same way. The addresses reserved for each device are summarised in a *memory map* which is shown in Figure 4.2. In it, the abbreviation 'k' is used to denote 1024 (2^{10}) bytes.

The smallest block of memory is 8K (2^{13}) so the lower 13 address bits are not needed to select any of the blocks. This leaves three address bits plus the signals /RD and /WR which are needed for controlling the RAM. The RAM needs two control inputs, the other five units need one each so a total of seven outputs must be supplied from our address decoder PROM. The PLE5P8 comes closest to this configuration with the following truth table:

/RD	/WR	A15	A14	A13	IO1	IO2	RAM	/W	P3	P2	P1	
X	X	L	L	L	H	H	H	H	H	H	L	– PROM1
X	X	L	L	H	H	H	H	H	H	L	H	– PROM2
X	X	L	H	L	H	H	H	H	L	H	H	– PROM3
L	H	H	L	X	H	H	L	H	H	H	H	– READ RAM
H	L	H	L	X	H	H	L	L	H	H	H	– WRITE RAM
X	X	H	H	L	L	H	H	H	H	H	H	– IO1
X	X	H	H	H	H	L	H	H	H	H	H	– IO2

From this truth table the following address table may be derived:

	0	1	2	3	4	5	6	7	8	9	A	B	C	D	E	F
0000	7E	7D	7B	7F	7F	7F	3F	5F	7E	7D	7B	7F	6F	6F	3F	5F
0001	7E	7D	7B	7F	67	67	3F	5F	7E	7D	7B	7F	7F	7F	3F	5F

The format used above is called *Hex ASCII* because the 'H' and 'L' patterns are converted to the hexadecimal equivalent numbers of address and data. ASCII is the American Standard Code for Information Interchange and is commonly used for sending data between electronic equipments. The active level for all the enable inputs is assumed to be LOW.

4.1.5.2 *Code converters and look-up tables*

A particularly suitable use of PROMs is for a device known as a look-up table. Mathematical tables of logarithms, trigonometric functions, etc. are well known, their use being to convert one number into another number which is a function of it. Although it is usually possible to devise an algorithm to

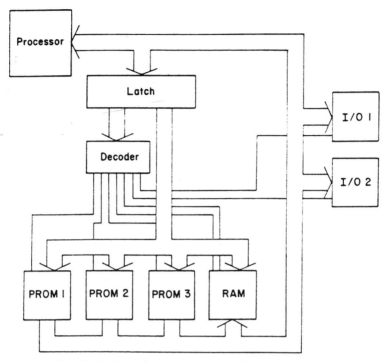

Fig. 4.1 Address decoder used in a microprocessor system.

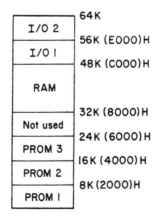

Fig. 4.2 Memory map for microprocessor system.

calculate the converted number this is often a fairly lengthy process. The alternative is to load a PROM with this information.

The PROM table is constructed by making the address correspond to the number being converted and the output data to the 'answer'. This is illustrated in Figure 4.3. A particularly useful conversion is 'twos complement'; addition and subtraction are two simple arithmetic functions which are common in

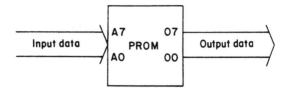

Fig. 4.3 PROM look-up table application.

electronic systems, as in other areas. The logic for these two operations is substantially different, but addition of the twos complement of a number is equivalent to subtracting that number. Thus if this function is readily available, the same circuitry may be used for subtraction and addition.

A code converter is just a specific example of a look-up table. Converting between decimal numbers and binary numbers is quite tedious but can be done very readily by a PROM. In all these cases it would be possible to work out the equivalent logic of the function being carried out by the PROM, and then create the function out of pure logic circuits. This will normally result in a far more complex solution.

4.2 PLAs

4.2.1 PLA architecture

4.2.1.1 Relationship with PROMs

The last section described how PROMs can be used to implement logic functions. Some of the practical limitations were discussed and we can now explore that subject further. The most notable feature of a PROM, when viewed as a logic device, is that every possible combination of input is present in the memory matrix. This makes PROMs very versatile but has some unpleasant consequences. Firstly, it means that there is often a high redundancy in the programmed information. In the address decoder example the logic function was described in seven lines of truth table, yet there were 32 lines of data programmed in the PROM. Secondly, adding an input to a PROM causes the size of the memory matrix to be doubled. This will have repercussions on cost and, to a lesser extent, on performance. Lastly, PROMs come with either four or eight outputs but logic systems often do not; thus output pins often remain unused, as one did in the address decoder example.

If we look again at the internal structure of a PROM it can be seen that the row decoder covers every possible input combination. Figure 4.4 shows a 3-line to 8-output decoder, as might be used in a 8 × 4 PROM, if such a device existed. The selection of any particular row depends on which of the six lines carrying the addresses and their complements are connected to the AND gate

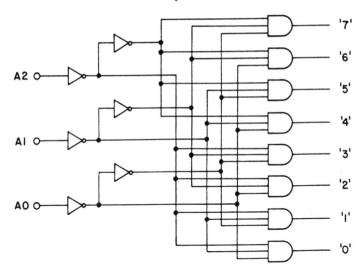

Fig. 4.4 PROM input address decoder.

driving that row. In a *programmable logic array*, or PLA, these connections are made programmable and the number of AND gates is reduced. Thus, instead of being fixed, the user decides which combination of input signals selects each AND gate. A typical structure is shown in Figure 4.5.

A PLA then has two programmable sections; the part described above, which connects the inputs to the AND gates is called the *AND matrix*. The second part, already described as part of the PROM structure, determines which AND gates are ORed together to form the output function; it is called the *OR matrix*. PLAs differs from PROMs by having a programmable AND matrix, although both have a programmable OR matrix. We can also examine two features of PLA outputs which can make PLAs more adaptable than PROMs.

4.2.1.2 *Programmable output polarity*

As already stated, PLEs may be considered to have programmable output polarity because each input combination can set the output HIGH or LOW. This is a desirable feature in any device with a limited number of AND gates, as some logic functions will use fewer gates with the outputs defined HIGH (in a Karnaugh map), while others may use fewer with LOW outputs.

The way this is achieved in a PLA is shown in Figure 4.6. One input to an exclusive-OR gate is connected to ground via a fuse; referring back to the truth table in Section 2.2.1.3, it is clear that with this fuse intact, that is with one input always LOW, the output level is the same as the input level. If the fuse is blown an internal resistor takes this input to a HIGH level and the output level is now the complement of the input.

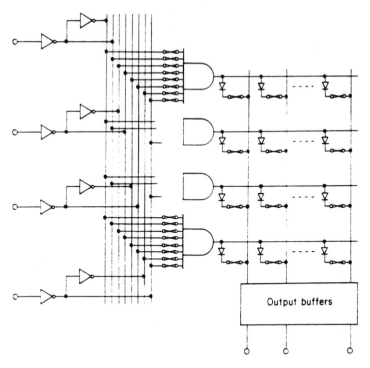

Fig. 4.5 Typical PLA structure.

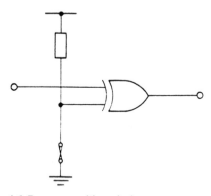

Fig. 4.6 Programmable polarity output structure.

In its non-inverting state the output is said to be *active-HIGH* because the OR gate output is left in the same sense, and will only be HIGH when one of the AND gates is active. Conversely, an inverting output is called *active-LOW*.

4.2.1.3 *Bidirectional I/O*

One of the stated disadvantages of PROMs was the restriction to four or eight outputs, which can prove wasteful of device pins. So far, we have not addressed this problem in relation to PLAs. By making use of the tri-state output structure, described earlier, some pins can be arranged to function either as inputs or outputs. This facility is illustrated in Figure 4.7. One AND

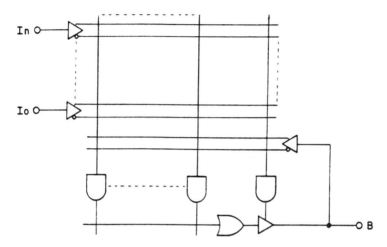

Fig. 4.7 Bidirectional output structure.

gate is dedicated to controlling the tri-state output, while the same pin is also connected to the AND matrix as an input. There are four options in the way in which this pin may now be used, depending on how the AND gate is programmed; these are:

- dedicated input pin – all fuses left intact
- dedicated output pin – all fuses blown
- controlled output pin – gate programmed with control logic
- output with feedback – output signal used in AND matrix

The control gates are often called *direction gates* because they define signal direction of the pin.

4.2.2 Designing with PLAs

4.2.2.1 *Configuring the AND matrix*

An AND gate can be made from a group of diodes connected by their anodes, so a component level structural diagram of a simple AND matrix is shown in Figure 4.8, the fuse indicating that the matrix is field programmable. For simplicity this is often drawn as illustrated in Figure 4.9, it being understood

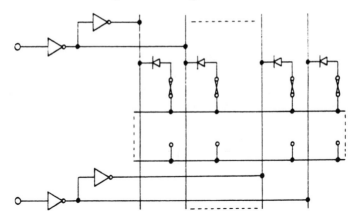

Fig. 4.8 AND matrix at component level.

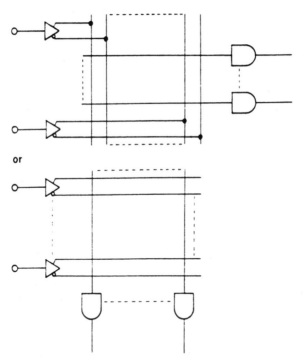

Fig. 4.9 Conventional representations of AND matrix.

that there is a diode and a fuse at each crossover. It is possible to map logic equations directly onto this matrix diagram. Because of the similarity between arithmetic and Boolean relationships, a single gate in the AND matrix is commonly called a *product term*.

Conventionally, an intact fuse is indicated by a 'X' or '.' at the crossover. Unprogrammed FPLAs have all fuses intact and should therefore have a 'X'

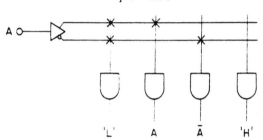

Fig. 4.10 Input connections to AND matrix.

at every crossover; this could be confusing, so they may be left out completely, with an 'X' in the AND gate symbol indicating an unprogrammed gate.

There are four possibilities for each input to be connected to each gate of the AND matrix, as shown in Figure 4.10. If both an input and its inverted signal are connected, we have the logic statement A*/A which is always LOW, therefore any gate with this included will inevitably be inactive. With the complement fuse blown, the true sense of the input (A) is included in the equation for that gate. If the inverted input is connected the complement (/A) will appear in the equation. When neither the input nor its complement is connected, that signal does not appear at all in the equation for that gate; this is the 'don't care' condition. If all the inputs are disconnected from an AND gate its output will be permanently HIGH, as may be seen from Figure 4.8.

4.2.2.2 Configuring the OR matrix

The OR matrix of a PLA is no different from that of a PROM. Figure 4.11 shows the component level structure and Figure 4.12 the conventional way of representing this. Each crossover is presumed to contain a fuse and a diode, connected by its cathode to the other diodes in the same gate. Again a 'X' or '.' is used to show that a signal from the AND matrix is connected to that OR gate and, therefore, included in that output function. It should be clear from the structural diagram that any OR gate with all its inputs disconnected will have a permanently LOW output.

Again by association with arithmetic, a gate in the OR matrix is often called a *sum term*.

4.2.3 Formal PLA methods

4.2.3.1 Truth table entry

We saw in the previous chapter that logic functions may be conveniently described by a truth table. This may be used as the basis for entering the information needed to programme a PLA. Most commercial programming equipment will accept a truth table as a direct entry method for PLAs, but it is worthwhile pointing out the relationship between the truth table data and the

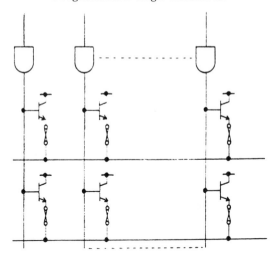

Fig. 4.11 OR matrix at component level.

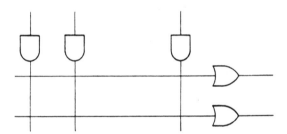

Fig. 4.12 Conventional representation of OR matrix.

fuses which are blown from that data. An 'H' in the truth table is equivalent to a true input, so this requires the complement fuse to be blown leaving the non-inverting connection intact. Conversely, an 'L' is interpreted by blowing the true fuse so that only the inverted input is connected.

The third possible entry in the truth table is '–'; this is the condition where the input is not included at all, so both fuses are blown in this case. The fourth possibility, both fuses being left intact, has the effect of disabling the gate to which they are connected, as we saw above. This condition is normally used only to specify direction terms in a truth table, when the symbol '0' is used.

We also need to specify whether the AND gate is connected to the output via the OR matrix, or not. An intact fuse means that the connection is made and this is shown by an 'A' in the appropriate output column. An AND gate which is not included in an output must have the fuse blown in the OR matrix; this is indicated by a '.' in the output column.

To illustrate these points, the truth table from Section 4.1.5.1 is reproduced in Figure 4.13 in the format normally used for PLA entry. A fictitious PLA with eight inputs and eight outputs has been assumed as the target device.

Inputs					Ouputs						
/RD	/WR	A15	A14	A13	IO1	IO2	RAM	/W	P3	P2	P1
–	–	L	L	L	.	.	.	.	.	.	A
–	–	L	L	H	.	.	.	.	.	A	.
–	–	L	H	L	.	.	.	.	A	.	.
L	H	H	L	–	.	.	A	.	.	.	.
H	L	H	L	–	.	.	A	A	.	.	.
–	–	H	H	L	A	.	.	.	.	.	.
–	–	H	H	H	.	A	.	.	.	.	.

Fig. 4.13 PLA truth table format.

4.2.3.2 Logic equation entry

Logic equations may be used as a source of programming data in one of two ways. Computer assistance for using logic equations is described in Chapter 8; it is also possible to convert equations to a truth table, by hand, and use this as the basis for data entry. True signals in the equation are entered into the table as 'H' and complemented signals as 'L'. Any signal not appearing in the equation is entered as '–'. Each line in the table corresponds to a single AND gate so the equations must be arranged into AND–OR format by using the rules specified in Chapter 2. If the same AND term, or product term, appears in more than one equation, it need be specified only once and connected to the necessary outputs by an 'A' in the appropriate column. We can illustrate the conversion process by deriving the equation for a D-type flip–flop. This, we recall, can be made from two D-latches in series. From Section 2.2.2.2 the equation for the first latch is:

$$Q1 = D * LE + Q1 * /LE + D * Q1$$

The second latch will use 'Q1' as its input and the complement of LE as its enable. This will ensure that the output from the first latch is passed through the second latch when LE changes from HIGH to LOW. The equation for the second latch, then, is:

$$Q2 = Q1 * /LE + Q2 * LE + Q1 * Q2$$

In the truth table, in Figure 4.14, the PLA is assumed to have four bi-directional I/O pins; a fifth signal has been added as an output enable, to show how this function is specified.

4.2.3.3 Karnaugh map analysis

We have already noted that the PROM structure contains every cell in the Karnaugh map, so it is possible to implement any combinational logic function in a PROM, provided that there are enough inputs. The power of the PLA is that inputs can be added without needing to increase the number of

Term	I/P /OE	LE	I/O-inputs D	Q1	Q2 Active level	LE −H	D H	I/O outputs Q1 H	Q2 H
00	–	H	H	–	–	.	.	A	.
01	–	–	H	H	–	.	.	A	.
02	–	L	–	H	–	.	.	A	A
03	–	–	–	H	H	.	.	.	A
04	–	H	–	–	H	.	.	.	A
D3	0	0	0	0	0				
D2	0	0	0	0	0				
D1	H	–	–	–	–				
D0	H	–	–	–	–				

Fig. 4.14 Truth table – D-type flip-flop.

AND gates. There is a price to pay for this and that is that there may not be enough AND gates to implement a given function. By mapping functions onto a Karnaugh map it may be possible to arrange the logic in such a way as to reduce the number of gates needed for the function.

It is not always possible to find a meaningful example to illustrate features which tend to occur in random logic situations. Some idea of what can be done is seen in the following example, although more obvious situations are likely to arise in real designs. The circuit in question has an output which equals the number of HIGHs on the input lines, as is shown in the following truth table:

I3	I2	I1	I0	O2	O1	O0	I3	I2	I1	I0	O2	O1	O0
L	L	L	L	0	0	0	H	L	L	L	0	0	1
L	L	L	H	0	0	1	H	L	L	H	0	1	0
L	L	H	L	0	0	1	H	L	H	L	0	1	0
L	L	H	H	0	1	0	H	L	H	H	0	1	1
L	H	L	L	0	0	1	H	H	L	L	0	1	0
L	H	L	H	0	1	0	H	H	L	H	0	1	1
L	H	H	L	0	1	0	H	H	H	L	0	1	1
L	H	H	H	0	1	1	H	H	H	H	1	0	0

The Karnaugh maps for the three output signals are shown in Figure 4.15. If this function is implemented in a PLA with no output inversions then 15 AND gates are needed, only one less than using a PROM! O0 needs eight gates, O1 six gates and O2 one gate. Even if the four cells which are 'H' in both O0 and O1 are shared a further six gates are still required to complete O1. However, if O1 is inverted one of its AND gates can be used for O2, four can be used in O0 and five more are needed to complete O0 and O1, a total of ten altogether.

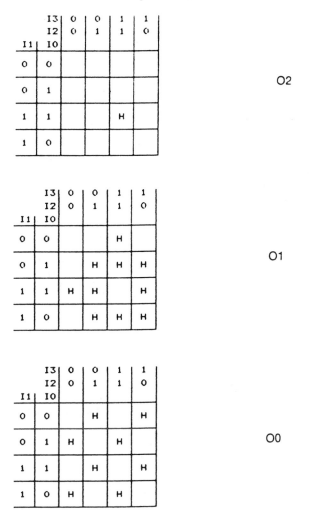

Fig. 4.15 Karnaugh maps – bit counting circuit.

This is not a startling improvement, partly due to the fact that we have limited the function size to four inputs for the sake of simplicity, but it illustrates the principle of logic minimisation in PLAs. That is to minimise the total number of gates by arranging for outputs to share gates wherever possible. We can also try to estimate the 'replacement value' of this solution over discrete logic. O2 clearly needs a single 4-input AND gate, while O0 would take three 2-input exclusive-OR gates. O1 is more complex however; it would use four 3-input NAND gates, two 4-input NAND gates and a 6-input NAND gate. This makes a total of five simple gate packages, which shows the amount of saving which can be made by using PLDs.

4.2.4 PLA availability

Most PLAs are sold under part numbers which bear no relationship to their
internal structure, unlike PLEs which we investigated earlier. In Table 3 the
'PEEL' devices are EECMOS, as may be deduced from their significantly
lower power consumption (specified at 10 MHz), the other devices are bipolar.
Current availability is as shown in Table 3.

Part number	*Input pins*	*I/O pins*	*Output pins*	*AND terms*	*Prop. delay*	*Supply current*
PLS100	16	—	8	48	50 ns	170 mA
PLS153	8	10	—	32	30 ns	155 mA
PLUS153	8	10	—	32	10 ns	200 mA
PEEL153/253	8	10	—	32	30 ns	45 mA
PLS173	12	10	—	32	30 ns	170 mA
PLUS173	12	10	—	32	10 ns	210 mA
PEEL173/273	12	10	—	32	30 ns	45 mA

Table 3   PLAs.

There are thus just three PLA architectures available although there is a
substantial range of performance within the limits indicated in Table 3. The
PEEL253 and PEEL273 have a similar structure to the 153/173 versions
except that their D-terms are in the OR-matrix; this allows an 'open-collector'
output to be obtained from any of the I/O pins.

The full circuit diagram of the PLS153 is shown in Figure 4.16.

4.3 PALs

4.3.1 PAL architecture

4.3.1.1 Derivation from PLAs

Where PLA stands for programmable logic array, PAL stands for programmable array logic. This may seem an artificial distinction introduced to
distinguish one manufacturer's product from an existing range, but it did
mark a radical departure from the PLE and PLA concepts established at that
time. If we take a historical look at the development of programmable logic we
can see how this came about. The first PROMs were introduced in about 1970
followed by a PLA, the PLS100, four years later. Although it was an extremely
powerful logic device, the PLA was not an immediate success. It was probably
too powerful to be appreciated fully by the designers of the time. Also, its
inflexible I/O structure limited the number of applications where it could
replace discrete logic effectively.

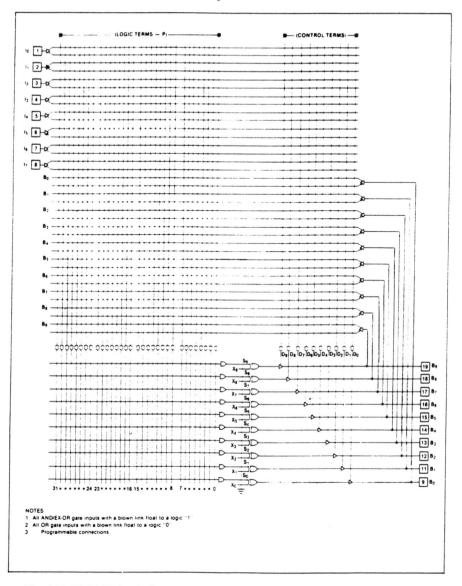

Fig. 4.16 PLS 153 circuit diagram *(reproduced by permission of Philips Semiconductors Ltd.)*.

Microprocessors, also introduced at about this time, were supposed to replace the discrete circuits used previously in logic designs; there was to be a revolution which would cut the use of TTL and CMOS integrated circuits to a fraction of their former number. It was found that sales of discrete logic circuits continued to grow in spite of all the predictions. The reason was that microprocessors needed logic circuits to interface with the other components used in their systems. Discrete logic became the 'glue' needed to stick systems together. Designers began to find that the glue was taking up more printed circuit board space than the microprocessors and their intelligent peripheral circuits. Although PROMs and PLAs could replace glue logic they were often an overkill solution, inasmuch as a large proportion of their logic power was wasted. A simpler solution was required and PALs stepped in to fill the bill.

Much of the PLAs power is derived from the fact that common AND gates can be shared between any of the outputs. Many simple logic functions do not overlap in this manner, however, so it is an unnecessary complication in many applications. In a PAL the AND gates are dedicated to a particular output by making the OR matrix fixed instead of programmable. Figure 4.17 shows how a simple PAL with four inputs, two outputs and eight AND gates would be constructed, while Figure 4.18 shows how the same effect can be obtained with a PLA. Restricting the OR matrix to fixed connections means that some of the flexibility of the PLA is lost, but this is compensated by being simpler to use.

4.3.1.2 PAL output structures

Although they have lost the programmable OR matrix, PALs do not suffer from any other drawbacks from the architecture standpoint. Indeed their simplicity makes them attractive to use for replacing discrete logic circuits.

The first PALs to be produced had an inflexible architecture. The outputs were fixed in number and in output polarity, and had few AND gates to share between them. This was fine for some applications, such as address decoders where the enabling inputs of the selected devices were all active-LOW, but could result in problems in other cases. Most of the PAL combinations are now available with programmable output polarity, just as the PLAs are. The other structure featured in some PLAs, bidirectional I/O pins, can also be found in some PALs.

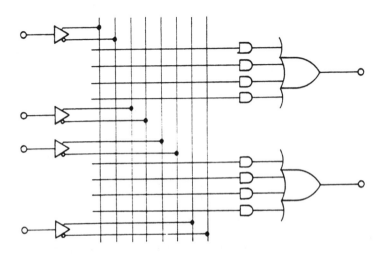

Fig. 4.17 Simplified PAL structure.

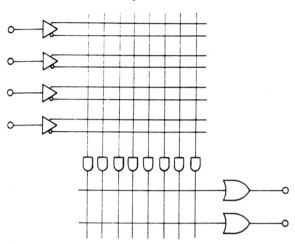

Fig. 4.18 PLA structure for comparison with PAL.

While PALs are very attractive for designing less complex circuits they do suffer from the limitation of restricting the number of AND gates which can be ORed together in each output. Although simple PALs have only seven or eight gates per output, attempts to cater for larger product term counts, by means of special output structures, seem to have been commercial failures. Either the demand is not there, or designers turn to PLAs or PLEs.

4.3.2 Using PALs

4.3.2.1 Design methods

There is very little difference between designing PLAs and designing PALs, except that for the latter truth table entry is not very usual. It comes down to establishing a method for describing the logic system in such a way that the appropriate fuses can be blown in whichever device is selected. The tools for assisting the designer will be covered in Chapter 8, here we are more concerned with seeing how the logic can be fitted into the various devices. It is the OR matrix which really governs the way in which logic minimisation is handled, and the Karnaugh map is the key to this.

The programmable OR matrix of the PLA made it politic to look for common AND functions in the maps of all the outputs. This approach will not help at all in PALs because the AND gates are allocated to a single output and cannot be shared. Thus the art in designing for PALs is to reduce the number of AND gates for each output. To fit the example of Figure 4.15 into a PAL would therefore require a PAL with the following properties:

Q2 – active HIGH output with 1 AND gate, or
  – active LOW output with 4 AND gates

Q1 – active HIGH output with 6 AND gates, or
   – active LOW output with 6 AND gates
Q0 – active HIGH output with 8 AND gates, or
   – active LOW output with 8 AND gates

The final choice of which device to use will depend on a number of factors, such as speed, power consumption and cost; the steps which need to be taken to achieve this will be discussed later in the book. The way in which the design information is converted to a finished device depends on the design tools which are available.

4.3.2.2   PAL nomenclature

Unlike PLAs, it is possible to deduce much about the structure of any PAL from its part number. PALs are numbered as PAL*mXn*, where '*m*' is the number of inputs to the AND gate matrix, '*n*' is the number of outputs and '*X*' defines the output structure. The common letters used for combinational PALs are:

H – active HIGH output
L – active LOW output
P – programmable polarity
N – address decoders

Most PALs are supplied as either 20-pin or 24-pin devices having, respectively, 18 or 22 logic connections. Thus if '*m*+*n*' comes to either 18 or 22 the device probably does not have bidirectional I/O pins and '*m*' is the number of device inputs, otherwise the number of inputs can be guessed by subtracting '*n*' from 18 or 22. An additional clue to the structure of a PAL comes from the fact that most PALs can be grouped into two classes of complexity, the simpler PALs having 16 or 20 AND gates in total with no bidirectional I/O, the others having 64 or 80 AND gates and bidirectional I/O. AND gates are normally spread evenly among the outputs so the number of AND gates per output can often be deduced.

4.3.2.3   Security fuse

A feature to be found in PALs but not in most PLAs is the security fuse. While the primary purpose of a PLD is to perform logic functions with fewer packages, there is a secondary benefit which could be even more useful in certain circumstances. The logic function contained in a PLD depends on the fuses which have been blown inside the device. While we have not yet discussed in detail the programming of PLAs and PALs, the methods are very similar to those for blowing PROMs. In other words they rely on internal circuitry which is in place for allowing the device to be programmed, and for reading which fuses have been blown.

Once a device has been programmed this extra circuitry is redundant. If it is disabled then it becomes impossible to read back the contents of the PLD and therefore its function. This makes it impossible, or at least very difficult, to copy. The security fuse is designed for just this purpose; once it has been blown in a PAL, that PAL cannot be forced to divulge its contents. It therefore becomes the ideal device to use in circuits which the designer wants to make difficult to copy, for commercial or other reasons. In some industries PALs are used to hold secret combinations, or to scramble signal paths, in order to make designs secure, so that competitors cannot make an identical equipment without having to invest themselves in design effort.

4.3.3 PAL availability

4.3.3.1 PAL families

As we stated above, PALs may be grouped in families according to their complexity. There are also PALs available within each group with different speed and power ratings, and different technologies. In our survey of combinational PALs we can therefore list the performance of devices at each end of the performance spectrum, low power and high speed. Low complexity PALs are those with 16 or 20 AND gates, medium complexity PALs have 64 or 80 gates.

PALs are fabricated in both bipolar and CMOS technologies but, in order to provide some choice of performance, bipolar PALs are available in different speed and power combinations. Because they have to cater for all their features to be fully used, the power consumption of both PALs and PLAs is often higher than the same circuit constructed from discrete logic. Just as discrete logic families are made with different levels of power consumption (e.g. ALS, AS and FAST), some ranges of PAL have been designed with lower supply currents than the standard family. This gives the designer the option of choosing devices with high speed, or low power if high speed is not required for the application. Families of half-power and quarter-power PALs are available.

The other option for low power is to use CMOS. The current consumption of CMOS is significantly lower at low speeds, but is comparable to bipolar devices at frequencies of about 10 MHz, or higher. This is because the internal nodes have to be charged and discharged more rapidly.

4.3.3.2 Combinational PAL summary

In Table 4, the various output configurations for a given I/O and gating architecture have been included on a single line:

All the AND term numbers, except for the low complexity PALs, include one direction term per output or I/O.

The 'special functions' are mainly PALs with only one product term per output for some or all of the outputs. These are intended for address decoding applications where only one input combination is active for each output. They may, of course, be used for any application where this condition applies.

Part number	Input pins	I/O pins	Output pins	AND terms	Low power p.d.	Low power I_{cc}	High speed p.d.	High speed I_{cc}
Low complexity PALs:—								
10H8 10L8	10	—	8	16	35	45	25	90
12H6 12L6	12	—	6	16	35	45	25	90
14H4 14L4	14	—	4	16	35	45	25	90
16H2 16L2	16	—	2	16	35	45	25	90
16C1	16	—	2(Q & /Q)	16	40	45	30	90
12L10	12	—	10	20	40	100	25	100
14L8	14	—	8	20	40	100	25	100
16L6	16	—	6	20	40	100	25	100
18L4	18	—	4	20	40	100	25	100
20L2	20	—	2	20	40	100	25	100
20C1	20	—	2(Q & /Q)	20	40	100	30	100
Intermediate PAL:—								
20L10	12	8	2	40	25	105	15	210
Medium Complexity PALs:—								
16L8	10	6	2	64	25	45	5	180
18P8	10	8	—	64	35	90	15	155
20L8	14	6	2	80	25	105	5	210
22P10	12	8	2	80	25	105	15	210
24L10	16	8	2	80	10	210	10	210
Special Functions:—								
16N8	10	6	2	16	5	180	5	180
18N8	10	8	—	8	6	180	6	180
85C508/9	16	—	8	8	7.5	48	7.5	48
48N22	36	12	10	73	7	420	7	420
7B336/8	12	—	8	16	6	180	6	180
7B337/9	12	—	8	32	7	180	7	180
ECL PALs:—								
10/10016P4	16	—	4	32	4	170	4	170
10/10016C4	16	—	8(Q & /Q)	32	2	220	2	220
10/10016P8	12	4	4	64	6	170	3	220

Table 4   Combinational PALs.

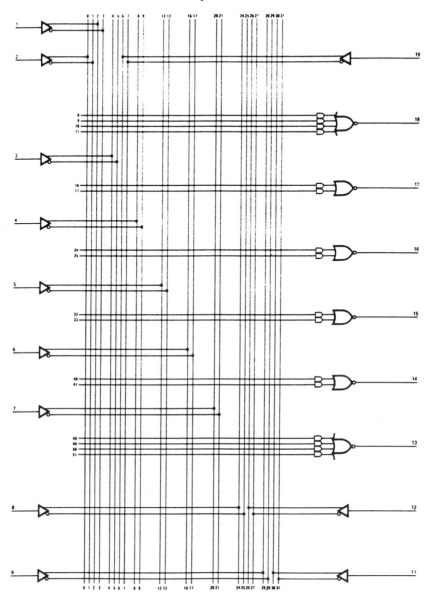

Fig. 4.19 PAL12L6 circuit diagram *(reproduced by permission of Advanced Micro Devices)*.

As an example of the actual PAL structures available Figure 4.19 shows the PAL12L6, Figure 4.20 the PAL20L8 and Figure 4.21 an ECL PAL the 10/10016P8.

4.4 REGISTERED PROMs

4.4.1 Application areas

4.4.1.1 Synchronisation

The simplest way to derive a sequential PLD from combinational devices is to feed the outputs into a D-type register. This structure is illustrated in Figure 4.22. The effect of this is to synchronise and store the outputs when the register receives an active clock edge. In every other respect the device acts in the same way as its combinational parent. A number of PROMs or PLEs are available with registered outputs, and this enhancement opens the door to more complex applications than are possible with the basic unregistered parts.

A common technique which is made possible by synchronous outputs is *pipelined operation*. This is a method of reducing the effective delay in systems where a number of operations are carried out in successive stages. An example of a simple pipelined system is the microprocessor instruction cycle, described in Section 2.2.2.1, although this uses a D-latch to store the address. The principle remains the same, however; the processor sends the address to the program memory at one stage of the cycle even though it does not require the result back until later in the cycle. Pipelining is often used in arithmetic operations to speed up throughput; Figure 4.23 shows a system for implementing Pythagoras' Theorem with PLEs.

Two methods are illustrated; the first uses combinational PLEs and, assuming a delay through each PLE of 30 ns, a result can be obtained every 90 ns when all the PLEs have settled. If a registered PLE with setup time of 30 ns, hold time of 0 ns and delay of 15 ns is used, a result can be obtained every 45 ns even though the data takes 135 ns to progress through the system. The benefit is obtained if a pipelined circuit forms part of a larger system; if the delays are comparable to those shown above then the system will be able to operate twice as fast as if unregistered PLEs were used.

4.4.1.2 State machines

We first described state machines in Section 2.3.2, where we saw that they can be constructed from a logic block driving a register whose outputs are fed back as further inputs to the logic. A registered PLE can thus be adapted to be a state machine by feeding some or all of the outputs back externally. The extended combination lock, which we used as an example in Section 2.3.3, fits well into a registered PLE as can be seen in Figure 4.24. The device required would be a PLE9R6; as with combinational devices the '9' and '6' refer to the

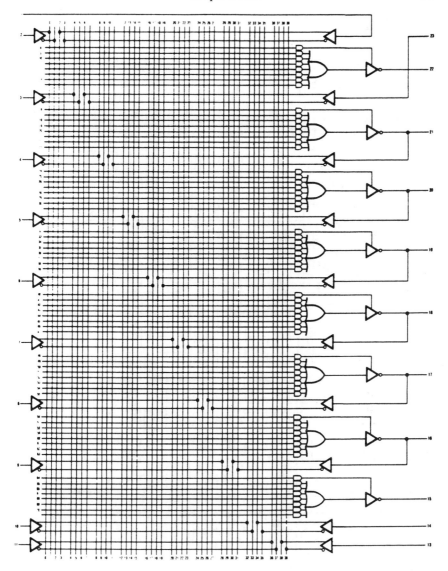

Fig. 4.20 PAL20L8 circuit diagram *(reproduced by permission of Advanced Micro Devices)*.

number of inputs and outputs respectively, while the 'R' denotes the fact that the PLE contains registers.

The logic equations for the device remain the same as in section 2.3.3.3. These need to be compiled into a format suitable for entry into a PLE, which is basically a memory device. To do this by hand is very tedious; in this example there are 512 possible input combinations and each has to be analysed to

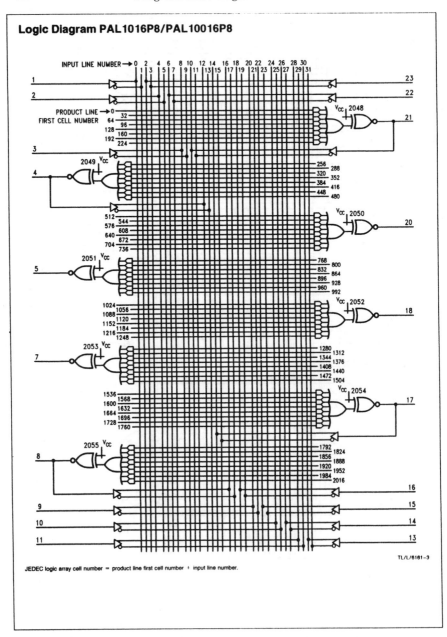

Logic Diagram PAL1016P8/PAL10016P8

TL/L/6161-3

JEDEC logic array cell number = product line first cell number + input line number.

Fig. 4.21 10/10016P8 ECL PAL circuit diagram *(reproduced by permission of National Semiconductor)*.

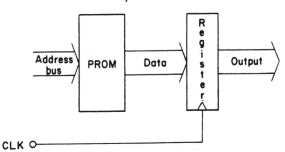

Fig. 4.22 Registered PROM.

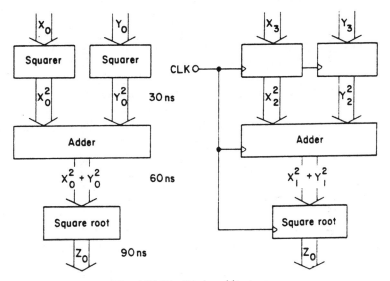

Fig. 4.23 Pipelined architecture.

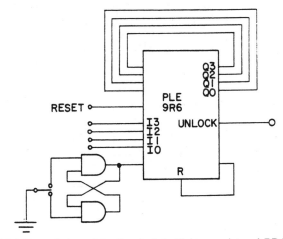

Fig. 4.24 Extended combination lock built in a registered PROM.

determine the output state for that combination. It may help if a Karnaugh map can be drawn for each output but these become unwieldy with more than about eight inputs. In practice, a computer may be used to generate the data; details of available programs will be given in a later chapter.

4.4.1.3 Diagnostic PROMs

We will cover the problems and requirements of testing in a later chapter, but there is a particular structure which makes systems more testable. This is called *level sensitive scan design*, usually shortened to LSSD. It is intended to solve the problems of observability and testability in complex systems like, for example, the pipelined Pythagoras calculator. If an incorrect answer is obtained from this system there is no way of knowing in which section the fault lies, without looking at the intermediate results. LSSD allows systems to be put into a test mode where the inputs to individual sections can be forced to a known state, and the resulting outputs observed. The way in which this can be done is shown in Figure 4.25.

In synchronous systems the output registers are given the ability to be loaded serially from a single external test point, by wiring all the registers into one long shift register. In test mode the data is fed in by clocking for as many cycles as there are cells in the registers. The system is then allowed to operate for a fixed number of cycles and the result clocked out serially. By comparing the result with the expected data the system can be tested and faults diagnosed with some precision. A family of *diagnostic PROMs* has been developed specifically for this type of application.

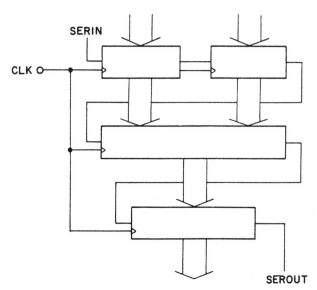

Fig. 4.25 LSSD system architecture.

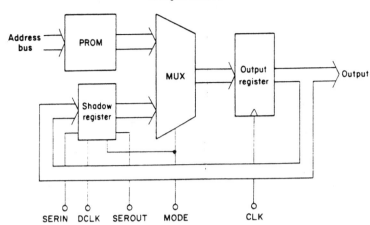

Fig. 4.26 Diagnostic PROM architecture.

Diagnostic PROMs, illustrated in Figure 4.26, include a second register called the *shadow register*. This can be loaded and emptied serially, and the contents transferred to and from the output register when the PROM is in test mode. These extra facilities use up four pins so the test facility is built in with very little overhead on device complexity.

4.4.2 Registered PROM availability

Registered PROMs are manufactured in both bipolar and CMOS technology; they have similar speed characteristics, but CMOS devices consume about half the power. The output enable may be applied in one of two ways; asynchronously when it acts in the same way as in an unregistered PROM, or synchronously when it is loaded into a flip-flop by the clock. Smaller registered PROMs include both options but the larger PROMs have only one or the other as there are not enough pins to allow both inputs. The asynchronous parts have the designation 'RA' whilst the synchronous parts are given 'RS'. The diagnostic PROMs are generally known by a manufacturer's part number rather than a PLE type number; available devices are listed in Table 5, with their switching specification in ns:

Part number	Memory size	I/P pins	O/P pins	Type of register	Switching spec.		
					tpd	setup	hold
PLE9R8	512×8	9	8	sync/async	15	30	0
PLE10R8	1024×8	10	8	sync/async	15	30	0
PLE11RA8	2048×8	11	8	async	15	30	0
PLE11RS8	2048×8	11	8	sync	15	30	0
PLE13R8	8192×8	13	8	sync/async	12	15	0

63DA441	1024 × 4	10	4	diag/async	18	35	0
63DA442	1024 × 4	10	4	diag/sync/async	18	35	0
63DA841	2048 × 4	11	4	diag/async	20	40	0
63DA1641	4096 × 4	12	4	diag/async	20	40	0
63DA1643	4096 × 4	12	4	diag/init	20	40	0
CY7C268	8192 × 8	13	3	diag/sync/async	12	15	0

Table 5   Registered PLEs.

This restricted range of devices is growing only slowly, if at all. The functions are of more interest as true memories, more so as memory size increases, so it is unlikely to excite much interest in logic designers.

4.5   REGISTERED PALs

4.5.1   Device structures

4.5.1.1   PLE restrictions

As in combinational logic, we find that registered PLEs are limited by the number of inputs which they can support. In applications such as pipelining this is not a disadvantage, because the purpose of the output register is to lock out the result of one stage of logic until the following stage is ready to accept it. The restriction on numbers of inputs can be a problem but no more than in combinational systems; after all a pipelined system is no more than a series of combinational logic blocks buffered from each other. Indeed a very common application is in arithmetic blocks where many AND terms are often needed and using a PAL or PLA is likely to be even more restricting.

Problems are far more likely to occur in state machine applications. In these cases some, or all, of the outputs must be fed back to the input so that the present state of the system can be incorporated into the state jump decision. In the worst case of a system needing eight outputs fed back, the largest registered PLE can accommodate only five more inputs. Furthermore, to generate the logic from even a simple state table is a very tedious task without the use of a computer, as we saw in the example which we discussed. In the same way that we derived the PLA and PAL structures from PROMs, we will see how registered PLAs and PALs are simpler to use and offer more scope for integrating state machines.

4.5.1.2 PALs with output registers

The simplest enhancement to a combinational PAL is the addition of an output register, as in Figure 4.27. In this class of device each output is loaded into a D-type flip-flop on an active clock edge, the registered output being fed back internally to the AND gate array. This structure is similar in principle to the registered PROM, except that the feedback is internal and does not use up inputs. The benefit of this has already been mentioned; in addition, the design is much simpler as many fewer AND gates need defining.

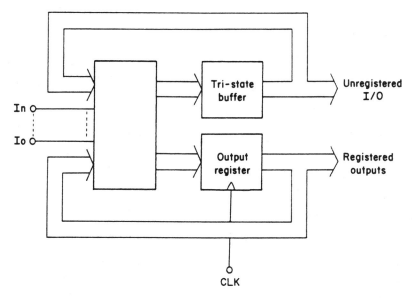

Fig. 4.27 Registered PAL.

In practice, a family of registered PALs is derived from one combinational PAL. For example, the PAL16L8 with registers becomes the PAL16R8, the 'R' indicating a registered part, and there are two other PALs with a similar structure; these are the PAL16R4 and PAL16R6. The PAL16R4 has four registered outputs and four bidirectional pins, while the PAL16R6 has six registered and two bidirectional pins. This allows state machines with up to 12 inputs to be built in the smallest package, because bidirectional pins can function as inputs. Alternatively, it is possible to mix combinational logic and sequential logic within one device.

4.5.2 Using registered PALs

4.5.2.1 Design example

We can see how a registered PAL can be used for the enhanced combination lock, described in Section 2.3.3. One problem with the equations of 2.3.3.3 is that they include terms such as:

$$/(I3 * /I2 * /I1 * /I0)$$

This could be expanded by DeMorgan's Laws into four OR gates but then Q3 and Q2 would need 12 AND gates each. Simple PALs have only eight AND gates per output so an alternative must be sought. If a PAL16R4 is used then the bidirectional pins can be used to form the complement functions, such as that above. Calling these intermediate functions C3, C2 and C1 respectively, the equations become:

$$/C3 = I3 * /I2 * /I1 * /I0$$
$$/C2 = /I3 * /I2 * /I1 * /I0$$
$$/C1 = I3 * /I2 * /I1 * I0$$

$$
\begin{aligned}
Q3 := &\ Q2 * /Q1 * /Q0 * C3 \\
+ &\ Q2 * /Q1 * Q0 * C2 \\
+ &\ Q2 * Q1 * /Q0 * C1
\end{aligned}
$$

$$
\begin{aligned}
Q2 := &\ Q2 \\
+ &\ /Q2 * /Q1 * /Q0 * C3 \\
+ &\ /Q2 * /Q1 * Q0 * C2 \\
+ &\ /Q2 * Q1 * /Q0 * C1
\end{aligned}
$$

$$
\begin{aligned}
Q1 := &\ /Q3 * /Q1 * Q0 * /C2 \\
+ &\ /Q3 * Q1 * /Q0 * /C1
\end{aligned}
$$

$$
\begin{aligned}
Q0 := &\ /Q3 * /Q1 * /Q0 * /C3 \\
+ &\ /Q3 * /Q2 * /Q1 * Q0 * C2 \\
+ &\ /Q3 * Q1 * /Q0 * /C1
\end{aligned}
$$

$$UNLOCK = Q1 * Q0$$

As the last equation shows, we can use the fourth bidirectional pin as an output to provide the UNLOCK signal. Care must be taken with the output polarity; the PAL16R4 has active LOW outputs, so the actual signals will be the complement of the above. Also we have not been able to include the asynchronous reset, as this feature is not included in the simple PALs.

Note that the ':=' symbol means that the equality is true only after the next active clock edge.

4.5.2.2 Hold and toggle methods

Registered PALs use D-type flip-flops almost exclusively. Referring back to our description of flip-flop types in Section 2.2.2.4 it is apparent that D-type flip-flops do not possess the inbuilt functions 'hold' and 'toggle'. These functions need to be incorporated in many sequential logic functions, such as counters and state machines in general. In order to incorporate these functions there are two possible approaches.

One way, which we followed in designing the enhanced combination lock, is to analyse the individual bits to define all the conditions which result in a HIGH being loaded. This means that each output is treated as a combinational logic circuit with the flip-flop appearing as an appendage. If we are to get the most out of registered PALs it is better to understand how the various sequential functions can be designed directly into the PAL. Loading a HIGH or LOW directly are straightforward as these functions exist implicitly in the D-type table; let us therefore turn our attention to the hold and toggle.

Hold, or don't change, means that if the output is already HIGH, a HIGH must be loaded by the next clock edge. If the output is LOW then it must remain LOW, which it will do if none of the AND gates feeding the flip-flop is true. This result can be achieved by feeding the output back to the input and gating it with the condition for holding. In other words an equation can be written as:

Q = HOLD CONDITION * Q

This has the desired effect because the equation is only true if Q is HIGH when the hold condition is present.

The requirement for toggling is just the opposite; a HIGH must become LOW after the clock while a LOW must change to HIGH. If we feed back the complement of the output then the ouput will toggle. The equation for this is:

Q = TOGGLE CONDITION * /Q

In this case, if the toggle condition is present, the equation is only true if the output is LOW.

4.5.2.3 Building a counter

We can apply these techniques to seeing how easily a counter may be defined as logic equations for a registered PAL. We can examine a 4-bit counter bit by bit and build the equations from the functions we require, instead of by constructing the state diagram. In addition to the basic counting function let us include a parallel load controlled by a 'LOAD' input, which enables data on four inputs to be put into the respective bits of the counter, and disables the count function while it is present.

The least significant bit (LSB) of the counter, Q0, has to toggle for every count and also has to be set HIGH when I0 is HIGH in the load condition. Thus the equation for Q0 is:

Q0: = /LOAD * /Q0 (toggle)
     + LOAD * I0 (load)

The next bit, Q1, toggles when Q0 is HIGH, holds when Q0 is LOW and loads when I1 is HIGH so the equations are:

Q1: = /LOAD * /Q1 * Q0 (toggle)
     + /LOAD * Q1 * /Q0 (hold)
     + LOAD * I1 (load)

The third bit toggles when both Q1 and Q0 are HIGH but must hold under all other count conditions, in other words when either Q1 or Q0 are LOW. Q2 is loaded from I2 giving the following equations:

$$Q2: = /LOAD * /Q2 * Q1 * Q0 \text{ (toggle)}$$
$$+ /LOAD * Q2 * /Q1 \text{ (hold)}$$
$$+ /LOAD * Q2 * /Q0 \text{ (hold)}$$
$$+ LOAD * I2 \text{ (load)}$$

The equations for the most significant bit (MSB) can be written down by following a similar argument:

$$Q3: = /LOAD * /Q3 * Q2 * Q1 * QO \text{ (toggle)}$$
$$+ /LOAD * Q3 * /Q2 \text{ (hold)}$$
$$+ /LOAD * Q3 * /Q1 \text{ (hold)}$$
$$+ /LOAD * Q3 * /Q0 \text{ (hold)}$$
$$+ LOAD * I3 \text{ (load)}$$

Similar reasoning will allow equations to be written directly for any systems which can be described in terms of the standard sequential functions.

4.5.2.4 *Exclusive-OR PALs*

One drawback of the counter design outlined in the previous section is that each extra counter bit requires an extra product term to define the hold condition. However, consider the equation:

$$D = Q : + : F$$

where F is some function to be defined. If F is HIGH, then $D = /Q$; whereas if F is LOW, then $D = Q$. Assuming that D is the input to a D-type flip-flop, the first condition will cause Q to toggle, the second will cause no change. Thus, if F is any of the toggle conditions in the counter equations above, the hold conditions become redundant.

Figure 4.28 shows a PAL structure with an exclusive-OR gate feeding the D-type flip-flop. This could be expanded to any number of bits without product terms limiting the size of counter which could be built.

The exclusive-OR PAL output will emulate a J–K flip-flop, as we can show in the following analysis:

consider $D = Q : + : (J * /Q + K * Q)$
if $J = 0$, $K = 0$ then $D = Q$
if $J = 0$, $K = 1$ then $D = 0$
if $J = 1$, $K = 0$ then $D = 1$
if $J = 1$, $K = 1$ then $D = /Q$

I will leave it as an exercise for the reader to expand the general equation, map it onto a Karnaugh map and show that it fulfils the conditions for a J–K flip-flop.

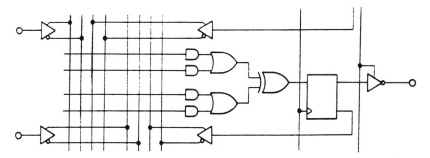

Fig. 4.28 Exclusive-OR PAL structure.

4.5.3 Registered PAL availability

4.5.3.1 Architecture and technology

The registered PALs described above may be considered as members of three families, each based on one of the combinational PALs. Thus, the PALs 16R4, 16R6 and 16R8 are derived from the PAL16L8, the 20Xxx family from the 20L10 and the 20Rxx PALs from the 20L8. The reason for treating them separately is to introduce registered PALs as an enhancement of combinational PALs; historically, all members of each family were introduced simultaneously.

The other architectural similarity is the pin-out of all three families. The clock input, which drives all output flip-flops in parallel, is always pin 1; the opposite corner pin is always an output enable for controlling the tri-state condition of each registered output. The latter feature means that in designs where the outputs are unconditionally enabled, it is easy to connect the output enable to ground, the pin on the opposite side of the DIL package. It should also be noted that there is no connection to the programmable array from these inputs, so connection will have to be made externally if either needs to serve as an input or output with respect to the rest of the logic.

All three families were originally fabricated in bipolar technology. Since then the speed has been improved by a factor of five over the basic specification. Devices are also available at the original speed but with a quarter the power consumption. CMOS parts are also available with slightly lower power still.

4.6.3.2 Registered PAL summary

Figure 4.29 shows the full circuit diagram of the PAL16R4. The PAL 'designation letters' used for simple registered PALs are:

R – 'plain' registered output (usually active-LOW)
X – exclusive-OR registered output

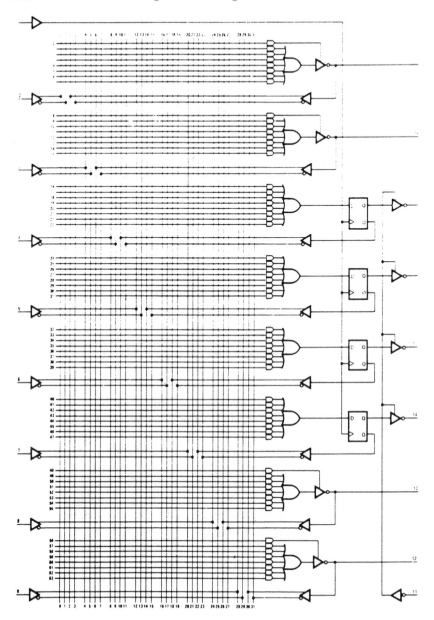

Fig. 4.29 PAL16R4 circuit diagram *(reproduced by permission of Advanced Micro Devices)*.

Some of the PALs listed in Table 6 are also available in 'RP' versions, that is registered with programmable polarity, although not over the full specification range.

Part number	Input pins	I/O pins	Reg. pins	AND terms	Low power $f._{max}$	I_{cc}	High speed $f._{max}$	I_{cc}
16L8 Family								
16R4	8	4	4	64	28	45	115	180
16R6	8	2	6	64	28	45	115	180
16R8	8	—	8	64	28	45	115	180
20L8 Family								
20R4	12	4	4	64	28	105	74	210
20R6	12	2	6	64	28	105	74	210
20R8	12	—	8	64	28	105	74	210
20L10 Family								
20X4	10	6	4	40	25	180	25	180
20X8	10	2	8	40	25	180	25	180
20X10	10	—	10	40	25	180	25	180
ECL PALs								
10/10016RM4A	12	—	4	32	200	240	200	240
10/10016RD8	8	—	8	64	117	280	117	280

Table 6 Registered PALs.

4.7 EXAMPLES

4.1(a) Draw up a PROM (PLE) truth table for a 3-input decoder, as described in Section 2.2.1.2.

(b) Do the same as (a), but in PLA format.

(c) Which simple PAL could also be used for this function?

4.2 Repeat the above exercise for the full adder of example 2.2. What advantage of PLAs over PALs does this illustrate?

4.3 Noting that registered PROMs and PALs contain only D-type flip-flops, show how each could be used to build a divide-by-five counter (as in example 2.4).

4.4 Complete the Boolean transformation of Section 4.5.2.4, showing how an exclusive-OR PAL can emulate J–K flip-flops.

Chapter 5
Complex PLDs

5.1 ENHANCED PALs

5.1.1 Generic logic

5.1.1.1 Output macrocells

So far, all the PLD structures we have looked at have had a fixed architecture; that is, each device has had a well defined number of combinational and/or registered outputs although many had some flexibility by virtue of the bidirectional I/O pin design. In the mid 1980s the first in a new class of PLD was introduced. As well as programmable cells in the logic path, they had programmable features in the device structure itself. We can call these, somewhat arbitrarily, complex PLDs and those with a fixed structure simple PLDs.

The key to programmable architecture is the *output macrocell*. Among the first devices with an output macrocell were GALs (generic array logic), which could emulate all simple PALs with eight or less outputs.

The circuit diagram of the GAL output macrocell is shown in Figure 5.1. As well as an output flip-flop and programmable polarity gate there are four multiplexers. These route the signals for the tri-state control, direct output, feedback and the 'eighth product term'. The desired signal path is selected by architecture cells which may be programmed in the same way as the logic cells. There are two global cells, called SYN and AC0, which define whether any registered outputs can be used and whether the GAL will emulate a small PAL (e.g. 10L8, 16L6 etc.), or a 16L8/20L8 type of PAL. In addition, each macrocell has an individual architecture cell, AC1, which defines whether the pin is an input or output, in small PAL mode, or registered or combinational in 16L8/20L8 mode.

This scheme allows the 20-pin GAL, the GAL16V8, to emulate any of the 20-pin PALs described in Chapter 4 (except the PAL16C1), and the 24-pin GAL20V8 to emulate any of the 24-pin PALs, except PAL12L10 and PAL20C1. Note that the feedback multiplexer needs inputs from two adjacent stages to cope with the functions of the 'corner pins'. In registered PALs they are dedicated to the clock and output enable, while in combinational PALs they are plain inputs; the small PALs and 16L8/20L8 also connect the inputs into the logic array in a different fashion.

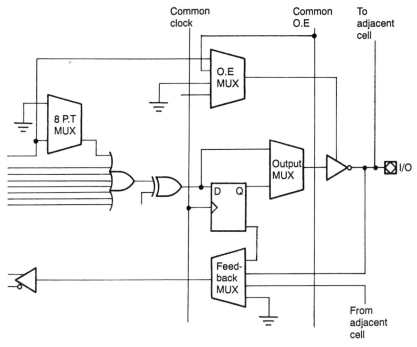

Fig. 5.1 GAL output macrocell.

5.1.1.2 *Other generic solutions*

While GALs may be considered to be the standard solution to simple PAL emulation, they are by no means the only solution. There are many variations on the macrocell theme and they have given rise to several alternatives to the GAL. They do have some features in common, aimed chiefly at simplifying the GAL architecture. Most have a dedicated product term for the output enable, plus eight logic terms per output, and they usually have permanent connections for the corner pins into the logic array.

Each macrocell can now function with only two, or possibly three, multiplexers, with selection being performed by two architecture cells per macrocell. With the product term allocation and output enable fixed, only the logic output and feedback signals use multiplexers. One of the most powerful of the generic PALs is the 22V10. The macrocell and block diagram are shown in Figures 5.2 and 5.3.

Apart from the features already noted, the 22V10 has a common reset and preset for its register bank, and a variable product term distribution. Two of the outputs have the usual eight terms, but two have ten terms, two have 12 and so on up to 16. This is an attempt to overcome problems of product term limitations which can affect PALs in some counter and other state machine applications.

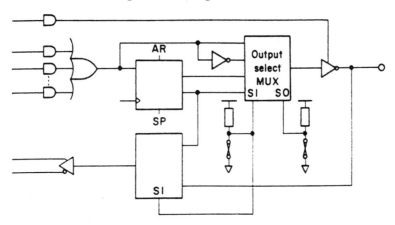

Fig. 5.2 PAL22V10 output macrocell.

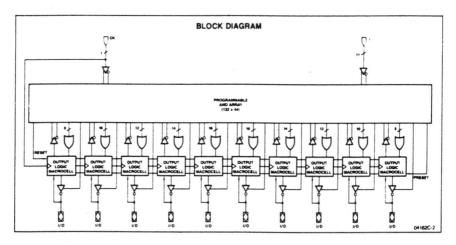

Fig. 5.3 PAL22V10 block diagram *(reproduced by permission of Advanced Micro Devices).*

5.1.1.3 Using generic PALs

In order to be successful at replacing simple PALs, the process must be semi-automatic. At first sight this does not appear to be the case. To configure a GAL, the eighteen architecture cells must be set and, in the case of the small PALs, the fuse map redrawn to take the larger fuse array of the GAL into account. There are two ways round this.

Although we have not yet discussed design and programming methods, we can assume that logic compilers and device programming machines exist to perform these functions. If a logic compiler has been used to create the fuse map for a simple PAL, the same logic information may be recompiled with a

GAL as the target device. Alternatively, most device programmers will reconfigure PAL fuse maps internally to fit into a GAL. In either case, the architecture cells will be set automatically to the correct configuration.

In the case of other generic PALs, reconfiguration by the programmer is not usually available. A software solution must be used. This may be by respecifying the target device, as described above, or by using a program supplied by the device manufacturer to reconfigure fuse maps to suit their particular architecture.

As well as replacing existing PALs, generic PALs may be used to create new PAL structures. Any of the macrocells in any generic PAL may be defined as a registered or non-registered, active HIGH or active LOW output, or, in most cases, as an input. There are some pitfalls to be avoided; in many devices, once any macrocell has been set to a registered output, the corner pins may not be used as logic inputs. A potential area of confusion is the polarity of feedback signals. If we compare the GAL and 22V10 macrocells, we can see that the feedback from the registered output is taken from the /Q flip-flop output. The programmable inversion, though, occurs before the flip-flop in the GAL but after it in the 22V10; and there is an inverting buffer between the macrocell and output pin. This confusion is resolved automatically if a logic compiler is used for the design, but it is as well to be aware of the difference between the two.

5.1.1.4 Zero-power PALs

With few exceptions, generic PALs are fabricated in CMOS technology. This allows the extra circuitry associated with the macrocells to be added with very little penalty in terms of chip area and power consumption. Examination of their data sheets reveals that the total supply current is still tens of milliamps rather than the microamps one normally expects from CMOS circuits. The reason for this is that, while the peripheral buffers and macrocell circuits are the usual low power CMOS, the logic array itself is not. A 'zero-power' CMOS gate is restricted to about four inputs, because of the stacking of transistors described in Chapter 1. The logic array has to be made from NMOS gates, each of which consumes a small standing current.

A few CMOS PALs are now available with zero power. These still use an NMOS array, but this is only powered up when a transition is detected at any input. The signal which enables the array power also enables output latches; these store the output data when the array is powered down. The schematic of a typical zero-power PAL is shown in Figure 5.4.

5.1.1.5 High output drive

So far, we have not looked at how PLDs can be interfaced to other parts of the circuit where they are being used. In general, they meet the current and voltage criteria of standard TTL families. This means that they will drive ten or more standard loads, and present no more than one standard load at their input. These interfacing criteria were designed to be adequate for connecting logic elements at up to 80 Mhz with 'reasonable' length connections on a printed

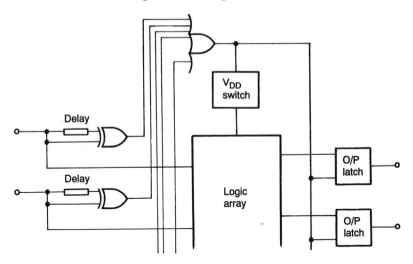

Fig. 5.4 Zero power PAL schematic.

circuit board. However, they may be found wanting in applications where high capacitance loads are found.

Examples of high capacitance loads are processor buses, motherboards and external cables. Discrete buffer circuits exist in the various TTL families to cope with these, so one option is to connect a PLD output via one of these to the load. Some PALs are available now with drive up to 64 mA (compared with the usual 16 mA), enabling a saving to be made in board space and supply current in high output drive applications.

These devices also feature input hysteresis, for signal noise filtering, so that they are suitable for bus interfacing, where slow signal edges are common, and for electrically noisy environments. Extra V_{cc} and ground pins are also included to help reduce supply line noise caused by high switching currents.

5.1.2 Asynchronous PALs

5.1.2.1 *Drawbacks of traditional structure*

Many circuits designed in standard logic rely on flip-flops clocked from different sources for their basic operation. To design a PLD to replace these circuits means either building the flip-flops from discrete AND–OR gates in a combinational PLD, or redesigning the function to build it from a synchronously clocked PLD. Other designs might need some flip-flops to be reset and others left unchanged by a reset signal. This again is a problem, as most PLDs have a universal set and reset which applies to all flip-flops, or none.

An example of an asynchronous circuit is shown in Figure 5.5. It will generate a waveform which is HIGH for 5/2 clock pulses and LOW for 3/2

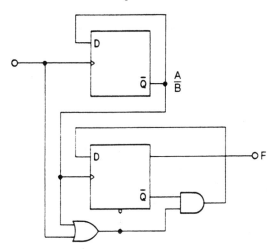

Fig. 5.5 Waveform generator with 5/3 mark/space ratio.

pulses. To make this function from a standard registered PLD would require a clock of double the frequency and an extra flip-flop.

5.1.2.2 The asynchronous PAL cell

In order to produce a range of PALs with a more flexible structure, including asynchronous clocking and individual flip-flop control, an asynchronous macrocell has been developed. The macrocell, which is shown in Figure 5.6, contains eight AND terms, but only four of these are combined to form a standard logic group. The other four provide a clock, set, reset and output enable which are dedicated to the flip-flop and output from that cell. By taking set and reset both HIGH, a condition which is normally illegal, the register is by-passed to convert the cell into a combinational output. All the outputs are fed back to the AND array making them, in effect, bidirectional pins.

A device is constructed by putting together as many cells as will fit into the package, the limitation being the number of inputs and outputs. Note that each cell uses one input and one output. In addition to the individual cell features there is a global output enable and parallel load. Thus a 20-pin package can accommodate eight cells, and a 24-pin package ten cells.

The provision of dedicated and global output enables means that each bidirectional pin can be set as an input or an output, or controlled on an individual basis, while every output can also be turned off simultaneously. The latter feature is necessary when using the preload in testing to set the flip-flops to a known state, apart from any possible use in normal device operation.

A slightly different cell design is shown in Figure 5.7; this is used in the EPseries of PALs. There are ten AND terms associated with each I/O cell, eight of these are used for creating the logic function, one is a clear which takes the flip-flop LOW and the tenth can be used as either a clock or output enable.

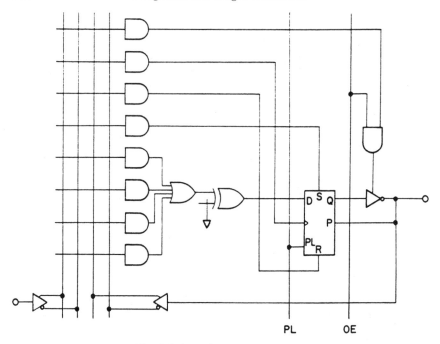

Fig. 5.6 Asynchronous macrocell.

A programmable multiplexer allows the clock to be taken from either the AND term or a global clock line giving the option of either individual clocking or a common clock. Similarly, the output enable can be either permanently ON or driven from the AND term. This arrangement is a little more restricting than the macrocell of Figure 5.6 as there is no preload facility and no global output enable, although these could be provided via the logic array if a common clock is being used.

The chief advantage of the EPseries output macrocell is its versatile choice of output type. The architecture bits may be programmed to provide almost any flip-flop configuration, or an unregistered output. Designs can be made with D-type, J–K, R–S or toggle flip-flops, and feedback selected from either the flip-flop or the pin for the D-type and toggle options. In J–K and R–S mode, the AND terms are shared between the two flip-flop inputs.

5.1.2.3 Design example

As half of the AND terms in the asynchronous macrocell are dedicated to specific functions, a particular convention is used to define the function in the logic equations. We can see how the convention is applied if we write the equations for the 5–3 mark–space waveform generator from Figure 5.5.

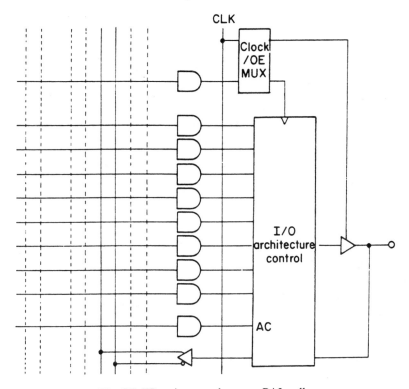

Fig. 5.7 EP-series asynchronous PAL cell.

$$B: = /B$$
$$B.CLK = A$$

$$F: = A * /F + /B * /F$$
$$F.CLK = /B$$
$$F.SET = /A * B$$

5.1.3 Buried flip-flops

5.1.3.1 *Macrocell structure*

The macrocells used for the enhanced PALs described so far have a significant drawback. If a macrocell is configured to be an input, the logic resource associated with that is lost. In the macrocell in Figure 5.8 there are two feedback paths to the logic array. If the output is disabled, allowing the pin to be used as a direct input, the feedback from the flip-flop is still available to the logic array. Note that this macrocell also includes an exclusive-OR product term; this allows J–K flip-flop emulation, as we described in Chapter 4.

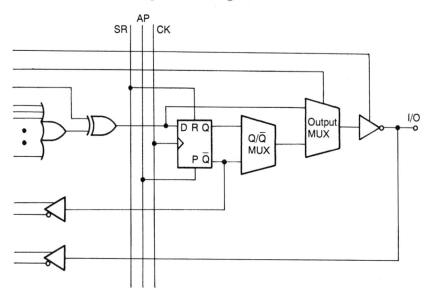

Fig. 5.8 Dual feedback macrocell.

The logic signal from the flip-flop is not observable from any device pin, when its output is disabled, so it is described as a *buried flip-flop*. Some PALs also include flip-flops which are permanently buried. They are aimed at Mealy machine implementation, since the device outputs can be a function of the buried register and the inputs.

5.1.3.2 *Device limitations*

Although registered PALs, and enhanced PALs in particular, can be used for implementing state machines, they do suffer from distinct limitations. If a state machine is built from J–K flip-flops, every transition requires one product term; if D-types are used, every LOW to HIGH transition requires a product term, as do cases where a HIGH is held. The PAL fixed OR structure means that, often, the same transition condition must be duplicated for several outputs. Because there are only between eight and sixteen product terms for each output in standard PALs, there is a danger that they will all be used before the design is complete.

Some amelioration may be achieved by logic minimisation, especially by judicious choosing of state numbering, but then there is a risk that the designer will lose visibility of the way in which the machine is designed to operate. As we shall see later, many of the logic compilers accept state transitions as an input; this may help the visibility but does not get around the fundamental problem of product term limitation.

5.1.4 Enhanced PAL availability

5.1.4.1 Summary Table

Table 7 summarises the enhanced PALs which are in production, or soon to appear, and are recommended for new designs. The term 'PAL' is used in its widest sense of being any device with a programmable AND array and fixed OR array.

Part number	Input pins	I/O cells	AND terms	Low power f_{max}	Low power I_{cc}	High speed f_{max}	High speed I_{cc}	Special features
GAL16V8	10	8	64	45	55	66	90	Generic macrocell
GAL16Z8	10	8	64	45	55	66	90	In circuit prog.
PEEL18CV8	10	8	74	33	35	50	105	Generic macrocell
PLD18G8	10	8	72	41	70	50	90	Generic macrocell
PAL23S8	9	8	135	33	210	—	—	Buried reg. Var AND
EP320/330	10	8	72	SB	0.15	100	75	Generic macrocell
85C220	10	8	72	NT	5	100	105	Generic macrocell
PLC18V8Z	10	8	74	21	0.1	—	—	Generic macrocell
PALCE16V8HD	10	8	64	50	90	—	—	64 mA Output drive
PLX464	10	8	98	28	80	—	—	64 mA O/P, var. AND
GAL20V8	14	8	64	45	55	66	90	Generic macrocell
85C224	14	8	72	NT	5	100	105	Generic macrocell
PEEL20CG10	12	10	92	37	55	50	105	Generic macrocell
PLDC20G10	12	10	90	33	55	45	70	Generic macrocell
PAL22V10	12	10	132	SB	0.1	111	190	Variable AND-terms
PAL22VP10	12	10	132	111	190	—	—	Variable AND-terms
PAL32VX10	12	10	152	22	180			22V10 superset/buried registers
PALCE29M16	5	16	188	33	100			Variable AND-terms/complex cell
PALCE24V10	16	10	80	37	90	—	—	Generic macrocell
PALCE26V12	14	12	150	40	105	—	—	Variable AND-terms
7C330	11	12	258	66	130			Var. AND/bur. regs/reg. inputs
7C332	14	12	192	43	130	—	—	Variable AND-terms
7B333	9	16	146	62	150	—	—	Buried registers
PAL16RA8	8	8	64	20	170	—	—	Asynchronous cell
PAL20RA10	10	10	80	45	75	—	—	Asynchronous cell
5AC312	10	12	192	40	90	—	—	Async./i/p latches
PAL22IP6	16	6	72	40	210			Async T/SR f-fs with 64 mA O/P
ATV750	12	10	170	55	120			Async dual f-f cell/var. P-trm
PALCE29MA16	5	16	178	30	100			Variable AND-terms/async. cell
EP610/630	4	16	160	SB	0.15	100	130	Asynchronous cell
85C060	4	16	160	SB	0.15	74	90	Asynchronous cell
7C331	13	12	192	23	120	27	130	Async./X-OR cell
10/10020EG8	12	8	90	(6ns)	285	—	—	Latched ECL cell
10/10020EV8	12	8	90	250	230	—	—	Registered ECL cell
10/10016ET6	10	6	48	(6ns)	240	—	—	ECL-TTL translator
10/10016TE6	10	6	48	(6ns)	220	—	—	TTL-ECL translator

Table 7 Enhanced PALs.

The numbers in the table need a certain amount of qualification. The numbers of input pins and I/O cells should be self-evident, however, the input pins do include any which can also be used as clocks or output enables, and may not be available for use as logic inputs in some configurations. The AND term count, likewise, includes those terms which are dedicated to output enable or common preset and clear. The manufacturer's data sheet needs to be consulted to assess the fine detail.

Performance figures relate to the lowest power and highest speeds offered, irrespective of manufacturer. Some 'shopping around' may be necessary to find the best compromise for a particular design. Where no 'high speed' figure is quoted, it usually means that all options have the same power consumption and the speed figure for best available selection is given. There could well be slower, cheaper devices for applications where speed is not so important. An 'SB' in the f_{max} column indicates a stand-by, or zero frequency figure; 'NT' means non-turbo mode.

Further explanation may be helpful for some of the more complex PALs, where 'special features' comments cannot tell the whole story. These are given on a device by device basis in the following sections.

Circuit diagrams of the GAL16V8 and PAL22V10 are shown in Figures 5.9 and 5.10.

5.1.4.2 GAL16Z8

This device is logically identical to the GAL16V8. Unlike the 16V8 it is intended for programming after it has been installed in a printed circuit board. It is in a larger package than the 16V8, 24 pins instead of 20, to accommodate the extra inputs needed for in-circuit programming. The chief benefit for the user is that, being electrically erasable, it can be reconfigured 'on the fly'.

Some care needs to be taken if frequent changes to its logic content are contemplated. Erasing and reprogramming EECMOS stresses the material and a limit of 10 000 cycles is normally put on this. A few cycles per day should give a life in excess of ten years, which is the data retention guarantee, but more frequent changes could reduce this significantly.

5.1.4.3 PAL23S8

The PAL23S8 is included with the 20-pin generic PALs in spite of being offered as a sequencer; this is because it does not contain all the features of sequencers to be discussed in Section 5.2. The features it does include are a buried register of six D-type flip-flops, an output register of four D-types and four macrocells. It also includes variable AND term distribution to ease the limitation problems associated with PAL type sequencers.

It can, therefore, be used for either a Mealy or Moore type of state machine, but not with the same ease of use or power as the sequencers we shall study later.

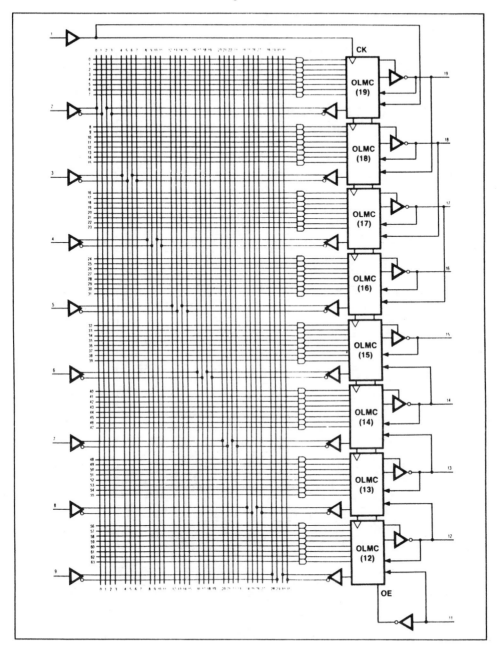

Fig. 5.9 GAL16V8 circuit diagram *(reproduced by permission of Lattice Semiconductors)*.

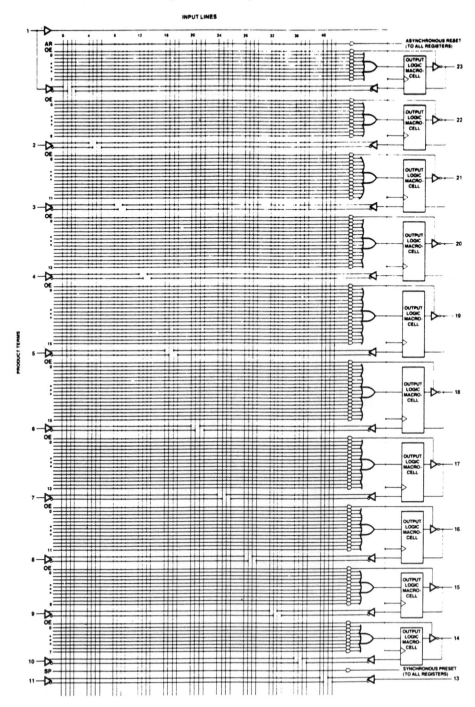

Fig. 5.10 PAL22V10 circuit diagram *(reproduced by permission of Advanced Micro Devices)*.

5.1.4.4 PAL32VX10

We have described this device as a 22V10 superset with buried registers. The AND term distribution is the same as the 22V10 but it has two important additions. Double feedback from each macrocell means that cells used as inputs can also act as buried registers, as shown in Section 5.1.3. One AND term per macrocell connects via an exclusive-OR with the other AND terms in order to emulate a J–K flip-flop.

5.1.4.5 PALCE29M(A)16

These devices are credited with complex macrocells. In fact there are two types of macrocell, eight with single feedback and eight with dual feedback to provide buried flip-flops. The active element in each macrocell can be either a flip-flop or a latch, and these can be used as an input register when the input mode is selected. The polarity of the active clock edge or latch level may be programmed and, in the PAL29M16, each macrocell can select its clock from one of two inputs. There are two options for output enable control, plus dedicated input and output settings; these are common enable from pin 11, or, in banks of four, from AND terms which are combined by an exclusive-OR gate.

In the 29MA16 the second clock option is an AND term, thus giving the possibility of asynchronous operation. Output enable control is from pin 11 or a single AND term, but on an individual macrocell basis. The asynchronous preset and clear are also per macrocell, instead of common as in the 29M16.

Both devices have preload and observability for all flip-flops/latches, even when buried, to improve testability, a subject we shall cover in a later chapter.

5.1.4.6 7C330

Although described as a synchronous state machine, the 7C330 has been included here, rather than as a PLS, because its registers are D-type and it has a PAL-type fixed OR array.

All the inputs and direct feedback paths from the I/O cells are registered, with a choice of two clocks, neither the same as the output register clock. The output register and buried register (four flip-flops) are D-type, as stated above, but have a single AND term exclusive-OR input to provide a J–K option with the appropriate feedback. The macrocells may be considered as pairs; each pair shares 32 product terms and an extra feedback buffer to the AND array.

Product term sharing within each pair is unequal, ranging from a 21/11 split to 17/15. As two terms in each macrocell are dedicated to output enable and exclusive-OR, respectively, there are at least 9 and as many as 19 terms available for logic among the various outputs. The shared feedback allows one member of each pair to be an input without sacrificing the use of the flip-flop, which becomes buried. The two buried register pairs also share product terms; in this case the usable splits are 19/11 and 17/13 as no output enable is required.

The schematic of one of the macrocells is shown in Figure 5.11.

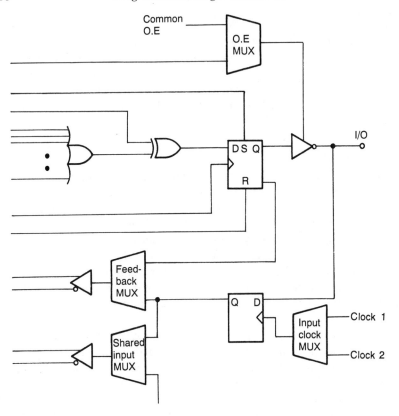

Fig. 5.11 7C330 macrocell.

5.1.4.7 7C332

The 7C332 might be considered as a combinational PAL, but it is included here as it has an input register, configurable as flip-flops, latches or buffers, and variable AND terms, from 9 to 19 per output. Two further AND terms provide an exclusive-OR connection for dynamic control of output polarity, and individual output enable control.

5.1.4.8 5AC312

This asynchronous device has eight latchable or registered inputs with either synchronous or asynchronous latching/clocking. In asynchronous mode the synchronous clock pins become available as normal inputs.

The macrocell, Figure 5.12, has pairs of product terms OR-ed together as synchronous clock, set, reset and output enable. The logic product terms are arranged in groups of four; any group not used in its own cell may be allocated to the adjacent cell, allowing for up to 16 terms in some cells at the expense of

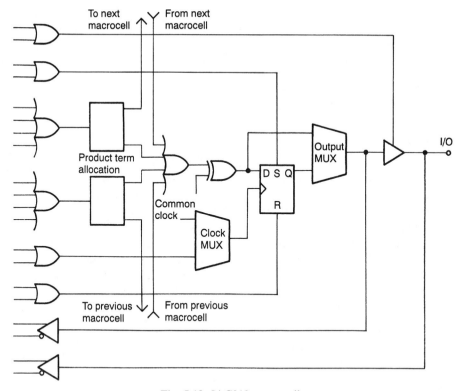

Fig. 5.12 5AC312 macrocell.

its neighbours. This allocation is controlled in 'rings' of six macrocells; unused terms from one ring cannot be allocated to a macrocell in the other ring.

Dual feedback, from the register and from the pin, give the possibility of buried registers in I/O pins used solely as inputs.

5.1.4.9 PAL22IP6

This PAL has two types of macrocell, three of each, but neither uses an external clock. One is a '2-T cell' with a pair of toggle inputs; either will cause the flip-flop to toggle independently, while simultaneous edges will cause no change. The second trio of cells contain an S–R flip-flop; an active edge into 'S' sets the output HIGH, conversely, driving 'R' sends it LOW. In addition, two of the outputs can drive 64 mA, the other four support 48 mA.

5.1.4.10 7C331

As well as providing an asynchronous clock, a single exclusive-OR term and varied AND term distribution, the 7C331 macrocell has a limited buried flip-flop capability. This is achieved in the same way as the 7C330, above, by sharing an extra feedback to the AND array. As with the 7C330, 32 product terms are shared between each pair; more terms are dedicated to functions within the macrocell, asynchronous clock, set and reset for input and output register as well as the output enable and exclusive-OR, so the logic term splits vary from 12/4 to 8/8.

There are no buried registers and the direct inputs are not registered, but otherwise the 7C331 has a similar architecture to the 7C330.

5.1.4.11   ATV750

The ATV750 also pairs off its flip-flops. In this case, however, one of each pair can be connected to the I/O pin whilst the other remains buried. Separate feedback from both flip-flops and from the I/O pin allows the pin to be used as an input, when the flip-flops become buried. The register is fixed as D-type only.

Product term sharing is also arranged somewhat differently; each flip-flop has either four or eight product terms allocated, but those allocated to the buried member of each pair can also be connected into the OR gate driving the other half.

The ATV750 macrocell is shown in Figure 5.13.

5.2   PROGRAMMABLE LOGIC SEQUENCERS

5.2.1   State machine implementation

5.2.1.1   Restrictions of PALs

We have seen how registered PALs can be used to implement state machines. The method was to draw the state diagram, derive the equations for the D-type register and then select the PAL that these equations will fit. This process is necessary because the output register uses D-type flip-flops which must be set whenever a HIGH output is specified. A LOW output is generated by the absence of an input signal. Many PALs have only eight AND gates per output so logic minimisation may be necessary if a complex state machine is being designed. The usual method of minimising logic is to use a Karnaugh map, but these must include the state bits as well as the inputs, so most complex systems will have too many inputs to make this method feasible.

The PAL structure does not therefore lend itself readily to complex state machines, although the use of design software can alleviate the problem if it performs logic minimisation. From the above discussion it may be surmised that the problems are due to the use of D-type flip-flops and the fixed OR-gate structure of PALs.

5.2.1.2   PLS flip-flops

Registered PLAs are called *programable logic sequencers* of PLSs, which emphasises their potential as state machine devices. In practice the PLS

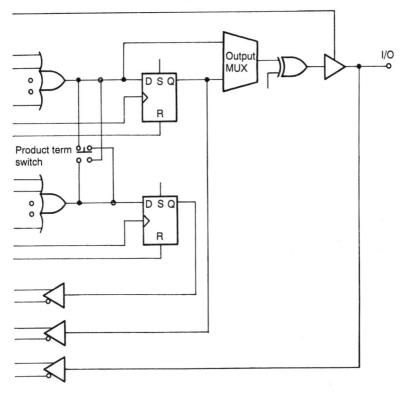

Fig. 5.13 ATV750 macrocell.

register uses either R–S flip-flops or J–K flip-flops. If we recall the truth table for the R–S device (Section 2.2.2.4), we see that three operations are possible; these are load HIGH, load LOW and hold. These operations make it possible to miss out the design stage of converting the state diagram to equations since the next state in any sequence can be loaded directly into the register, as the design example in Section 5.2.2.3 shows.

The one operation which cannot be carried out directly by an R–S flip-flop is to toggle. J–K flip-flops, on the other hand, include the toggle operation along with the three R–S modes. Most state machines, particularly those involving counting, can be made more economically with J–K flip-flops because of their larger number of operating modes. Perhaps more important is the fact that design is made more easy because the state transitions do not need transforming into AND–OR type equations.

5.2.1.3 *Benefits of the PLS structure*

The second disadvantage which we described with PALs was the fixed OR array which limits the number of AND gates available for each output. PLSs, like PLAs, contain a programmable OR array. The real benefit of this in state machines is that each jump condition needs only one AND gate which can set or reset every flip-flop. This becomes clear when the actual structures are described, with an example, in the next section.

5.2.2 PLSs with R–S flip-flops

5.2.2.1 Basic structure

Figure 5.14 shows the structure of the first PLS famly to use R–S flip-flops. The AND gate array and OR gate array have the same structure as a PLA, that is the true or complement of any of the inputs can be gated together and the resulting logic function OR-ed with any of the other AND functions to form an output function. In the case of the PLS these output functions are used to drive either the set or reset of the R–S flip-flops; this set of flip-flops is called the state register, as they contain the information as to the present state of the PLS.

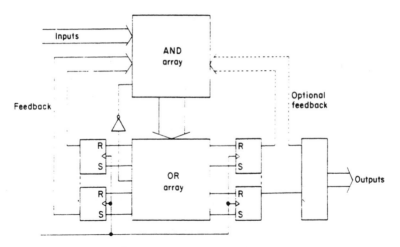

Fig. 5.14 PLS with R–S flip-flops.

The state register is divided into two parts, a buried register section which is fully fed back as inputs to the AND gate array, and an output register, only part of which is fed back. Six feedback signals allow for up to 64 states to be defined, which should be enough for a state machine with 20–24 I/Os. Those PLSs with more feedback signals give the designer the opportunity to define states in a way which may be more meaningful, or allow more efficient use of product terms.

This architecture is compatible with the Mealy type of state machine in which the output is not tied directly to the internal state of the machine but may, for example, depend on the input as well as the present state. An example would be a keyboard encoder in which the fed-back signals are used to perform the scanning and debounce functions; then, having detected a depressed key, would set the output to the code corresponding to the key.

If all the outputs are fed back to the AND array then the PLS behaves as a Moore machine, in which the outputs depend only on the present state of the machine. Most counters and simple controllers are examples of Moore machines.

Enhancements of this basic structure are discussed after the summary table in Section 5.2.4.

5.2.2.2 Complement term

One part of the PLS which has not appeared in the other logic devices described so far is the complement term. This is a single NOR gate whose inputs are any of the AND functions, and which is fed back to the AND array. As it is a NOR gate its output will be LOW if any of the AND gates driving it is HIGH; in which case it will cause any AND gate to which it is connected to be inactive. However, if all of its inputs are LOW its output will be HIGH and it will allow any gate to which it is connected to be active provided all the other inputs to that gate are true as well.

It may be used to define default conditions within the state diagram (see Figure 5.15). If a number of jumps are defined out of a given state a default jump can be defined if none of the other input conditions is true. This is done by connecting all the defined jumps to the complement term, and gating the output with the present state in a further AND gate. It is usually more efficient to do this than define all the other possibilities individually.

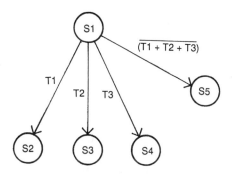

Fig. 5.15 Complement array.

By gating with the present state it is possible to use the complement term for more than one present state. If the register is in state 'A' then any AND gate which includes state 'B' must be inactive so the output of the complement term depends only on whether any of the defined jumps out of state 'A' is true. Even if its output is HIGH the AND gate including the complement term and state 'B' will be LOW and will not affect the outputs from the AND array.

More general use can be made of the complement term as a way of reducing the number of AND gates needed for all the jumps from a given state. When we were looking at the Karnaugh map in relation to combinational logic we saw that programmable outputs allowed us to group either '1's or '0's and choose the lower number of groups. If we draw the Karnaugh map for just the input conditions for each state jump we must group the '1's if we are to drive the flip-flops directly. However, if we use the complement term, we can group the '0's to give us the same function. This can be used as the solution if fewer AND gates are required, remembering that an extra AND gate is required to gate the complement with the present state.

The only drawback with this technique is that the extra feedback path introduces an extra delay into the logic path. It is not applicable, therefore, to designs where speed is at a premium. As an indication of the effect of this extra path, the setup time is increased by about 30 ns, while operating frequency can be reduced from 20 MHz to 12.5 MHz.

5.2.2.3 Design Example

The design method for PLSs is best explained by reference to a worked example; we can use the enhanced combination lock again as an illustration. As we can bypass the logic equation stage we can go back to the state diagram and translate this directly to a state table. Referring to Figure 2.27, where we originally introduced the state diagram, we can define all the state jumps directly as:

Inputs				Present state				Next state				O/P
I3	I2	I1	I0	Q3	Q2	Q1	Q0	Q3	Q2	Q1	Q0	UN
H	L	L	L	L	–	L	L	L	–	L	H	L
L	L	L	L	L	–	L	H	L	–	H	L	L
H	L	L	H	L	–	H	L	L	–	H	H	H

These three lines of state table define all the 'legal' jumps in the state diagram. If the three AND gates are also connected to the complement term we can use this to define the jumps for incorrect entries. As long as one of the input/present state combinations is true the complement term will have a LOW output. If an incorrect entry is made then none of the inputs will be HIGH so the complement output will be HIGH. If this is gated with the 'first mistake' bit, Q2, the incorrect entry will set either Q2, if this is the first mistake, or Q3 if it is the second.

The convention for incorporating the complement term into a state table needs some explanation. In principle it should appear in both the output and input side of the state table as it can take on either role. In practice it is given a separate column on the input side with an 'A' used to define it as an output (generate) and a '.' as an input (propagate). If it is not being used as either it should be programmed '–' (don't care) so that it does not interfere with normal operation. The full state table then appears as:

| | | *Inputs* | | | | *Present state* | | | | *Next state* | | | O/P |
C	I3	I2	I1	I0	Q3	Q2	Q1	Q0	Q3	Q2	Q1	Q0	UN
A	H	L	L	L	L	–	L	L	L	–	L	H	L
A	L	L	L	L	L	–	L	H	L	–	H	L	L
A	H	L	L	H	L	–	H	L	L	–	H	H	H
.	–	–	–	–	L	L	–	–	L	H	–	–	L
.	–	–	–	–	L	H	–	–	H	H	–	–	L

Note that in this design we have specified UN as the 'unlock' function and have generated this synchronously as a registered output which plays no part in the state jumps. PLSs in this family can be given an asynchronous preset which would set the register to the complement of what we require from the RESET we specified previously. The registers also power-up in a state with all HIGHs so a practical solution would need the state bits exactly the inverse of the way we have specified them. That this would not affect the operation of the circuit at all goes to show the versatility of this type of device. It is also remarkable that a circuit with nine states and 12 transitions can be specified in just five AND terms!

The other important benefit of this structure is that we were able to enter the states directly into the state table, which is a format accepted by most commercial programming equipment. This also allowed us to perform logic minimisation by inspection, by recognising that the 'legal' jumps are independent of the 'first mistake' bit, Q2, which can therefore be set as don't care. In a more complex system we might have to use one of the formal techniques already described if the logic will not fit into the PLS.

5.2.3 PLSs with J–K flip-flops

5.2.3.1 *The composite flip-flop*

The macrocell used by most of the PLSs with J–K flip-flops is shown in Figure 5.16. This 'composite flip-flop' includes many added features which allow it to act in more ways than just an output register.

The first addition is a programmable inverter driving the 'K' input from the 'J' input. Provided that the 'K' input is otherwise disabled, this feature makes the flip-flop behave in the same way as a D-type, for if 'J' is HIGH then 'K' is LOW so a HIGH is loaded, while a LOW on 'J' sends 'K' HIGH to load a LOW. This is the corollary to the exclusive-OR product term driving D-types; it enables the J–K to emulate D-types in those applications, such as registers and pipeline circuits, where the D-type is more efficient. The invertors are controlled by a separate fuse array and AND gate so that the type of flip-flop can be either preset or changed during circuit operation.

Also connected to the 'J' and 'K' inputs are the outputs from a true/complement buffer driven from the device outputs. This is the same as the register preload described in connection with some of the more complex

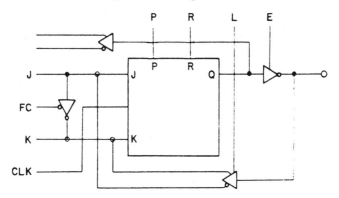

Fig. 5.16 Composite J–K flip-flop.

PALs. Preload should only be used when the outputs are disabled by the tri-state control to avoid contention on the output pins. As with PALs, this facility may be used to assist in testing or to ensure proper start-up of a state machine. Another possibility is that the output pins can be connected to a data bus, the register loaded with data which is modified according to the input conditions and loaded back onto the bus. The tri-state control operates after the feedback into the AND array so the PLS can be hidden from the bus until its data is ready to be read.

As well as these two novel features, the composite flip-flop has asynchronous preset and reset making it a very versatile circuit element.

5.2.3.2 PLS155 Family

The principle family of PLSs using this composite flip-flop macrocell is the four-member PLS155 group. They are, as listed in Table 8, the PLS155, PLS157, PLS159A and PLS179. Their common architecture is shown in Figure 5.17. It is, in effect, an extension of the PLS153 architecture. The unregistered section of the PLS is a reduced version of the PLS153 with a set of inputs and bidirectional pins, and an array of AND gates driving the programmable OR array. As with registered PALs, this makes it possible to mix registered and unregistered functions in the same device, or dedicate all the unregistered pins as inputs.

Unlike the R–S type PLSs, there is no buried register in the PLS155 family. All the flip-flops are connected to output pins and are also fed back to the AND gate array, so these PLSs can be used to build Moore-type state machines. The gates which drive the preset, reset and parallel load functions of the composite flip-flops are located alongside the tri-state control gates for the unregistered bidirectional pins. All these gates form a control section of the PLS.

The only control function which does not reside in this area is the flip-flop type control. This is the term which defines the behaviour of the flip-flop;

Part number	Input pins	I/O cells	Trans. terms	Bur. regs.	Speed $f_{.max}$	Power I_{cc}	F/F type	Other features
PLSCE105	16	8	48	6	25	100	RS	Preset/OE option
PLUS105	16	8	48	6	35	200	RS	Preset/OE option
PLUS405	16	8	64	8	38	225	RS/JK	Twin 4-bit regs.
PLS167	14	6	48	6	30	160	RS	Preset/OE option
PLS168	12	8	48	6	23	160	RS	Preset/OE option
PLS506	13	8	97	16	33	210	RS	Reg/Com O/P opt.
PSG507	13	8	80	8	33	230	RS	Int. 6-bit counter
PLS30S16	13	12	64	4	25	225	RS	4 dual f/b I/Os
PLS155	4	12	32	—	14	190	JK	4 Reg. 8 comb I/Os
PLS157	4	12	32	—	14	190	JK	6 Reg. 6 comb I/Os
PLS159A	4	12	32	—	18	190	JK	8 Reg. 4 comb I/Os
PLS179	8	12	32	—	18	210	JK	8 Reg. 4 comb I/Os
PLC42VA12	8	12	64	—	15	120	JK	10 async. dual f/b
PLC415	16	8	64	8	13	80	JK	Twin 4-bit regs.
GAL6001	11	10	64	8	27	150	D/E	Reg. i/ps – no c/a
XL78C800	12	10	32	—	22	35	JK	Foldback NOR arch.
7C361	8	14	32	—	125	150	D	Shift reg. arch.

Table 8   PLSs.

whether it acts as a J–K or as a D-type. Each of the J-to-K inverters has a fused connection to an AND-gate output called the flip-flop control term. It is logically connected so that the inverter is enabled by a LOW on its control input. Because the AND gate has a LOW output in its unblown state it follows that the flip-flops will behave as D-types if all the fuses are left intact. To achieve J–K operation it is necessary to ensure a HIGH on the control input. This will be the case if the control fuse is blown, or if the flip-flop control term is at a HIGH level itself.

The other features which this family shares with the R–S family are the complement term and a tri-state enable for the registered outputs. We have already dealt with the use of the complement term but two more points are worth making regarding its use in this family. Firstly, it may be used with the unregistered outputs as an internal feedback to make, for example, an exclusive-OR function without using up outputs. Also, its output may be connected to the control section. A possible use might be to force a reset if an illegal state is detected.

The tri-state enable has to be used in conjunction with the parallel load function. If this function is not being used and tri-stating is not required it is possible to permanently enable the outputs by blowing the appropriate fuse, as detailed in the device data sheets. One possibility which should not be overlooked is that the registered outputs could be used as registered inputs instead by leaving the outputs permanently tri-stated. The device can be used

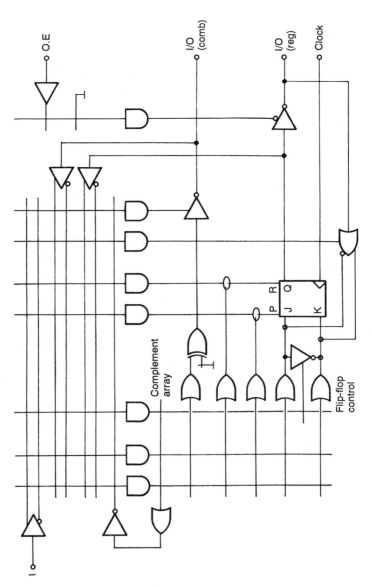

Fig. 5.17 PLS155 family architecture.

as a PLA with registered inputs by treating the bidirectional pins as the only outputs. This technique is used when interfacing asynchronous signals to a synchronous system; using a PLS has the advantage that logic functions can be included in this *staticiser*, as it is called.

5.2.3.3 *Design example*

Once again we can best illustrate the design procedure by means of an example. Let us again build a 4-bit counter but this time extend the function so that we have the option of a binary count or Gray code counting. We have already encountered Gray code in making Karnaugh maps; in Gray code only one output at a time toggles so possible problems due to dynamic hazards are avoided when decoding the count. Firstly, though, we will see how to build a binary counter from the J–K family, and then design the additional Gray code counter.

In the discussion on registered PALs in Section 4.5.2.3 we saw how to build a counter with D-type flip-flops, and how the number of terms for each bit escalated as we progressed up the counter bits. This was due primarily to the need to define the hold conditions. J–K flip-flops have the hold condition built in so, without analysing the situation formally, it seems likely that J–Ks are going to provide a simpler solution than D-types. In fact, all we have to define is that each bit toggles when all the lower order bits are HIGH. A J–K toggles when $J = K = HIGH$ so we need to program the following equations:

$$Q0.J = BIN$$
$$Q0.K = BIN$$

$$Q1.J = BIN * Q0$$
$$Q1.K = BIN * Q0$$

$$Q2.J = BIN * Q0 * Q1$$
$$Q2.K = BIN * Q0 * Q1$$

$$Q3.J = BIN * Q0 * Q1 * Q2$$
$$Q3.K = BIN * Q0 * Q1 * Q2$$

BIN is a signal which we use to indicate that we are counting in binary. In order to convert them into a state table we must see how the flip-flop inputs are specified. Each flip-flop has two inputs, 'J' and 'K', and we can write up a truth table relating fuse conditions to the functions and table entry convention:

'J'	'K'	Code	Function
intact	intact	0	toggle
intact	blown	H	load HIGH
blown	intact	L	load LOW
blown	blown	–	hold

Using these conventional symbols we can generate the following state table:

| | *Inputs* | | | | *Outputs* | | | |
BIN	Q3	Q2	Q1	Q0	Q3	Q2	Q1	Q0
H	–	–	–	–	–	–	–	0
H	–	–	–	H	–	–	0	–
H	–	–	H	H	–	0	–	–
H	–	H	H	H	0	–	–	–

We may now design the Gray code half of the counter. The sequence of counting is shown in Figure 5.18; one way to put this into a PLS would be to enter each transition as a separate AND term. This would use 16 AND gates, so let us investigate the possibility of reducing this number by examining the transitions of each bit in turn. Since we have the option of using D-types or J–K flip-flops we can draw the Karnaugh maps for both loading HIGHs and toggling. The results are shown in Figure 5.19. The number of AND gates required for each map is indicated by the map.

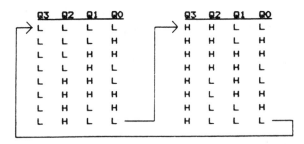

Fig. 5.18 Gray code count sequence.

Not surprisingly, using J–K flip-flops exclusively would take 16 gates because the property of Gray code is that the bits toggle one at a time when counting. The appearance of the maps suggests that the exclusive-OR function would be useful; for example we can write Q0 as:

/Q0: = Q3 : +: Q2 : +: Q1

However, this would take more gates to build than merely entering the four gates directly from the map, even if we use the complement term. On this basis we could build the Gray counter from 13 gates using D-type flip-flops, by reference to the Karnaugh maps. More careful analysis shows that we can save two more gates by using J–Ks for Q3 and Q2, and D-types for Q1 and Q0. The equations for this section are, therefore:

$$Q3.J = /BIN * /Q3 * Q2 * /Q1 * /Q0$$
$$+ /BIN * Q3 * /Q2 * /Q1 * /Q0$$
$$Q3.K = /BIN * /Q3 * Q2 * /Q1 * /Q0$$
$$+ /BIN * Q3 * /Q2 * /Q1 * /Q0$$

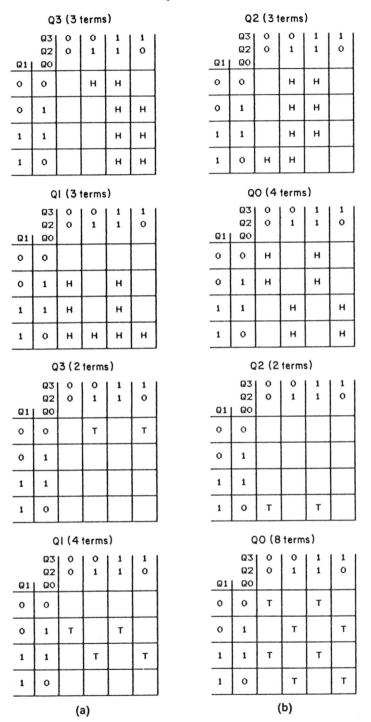

Fig. 5.19 Karnaugh maps – Gray code counter. (a) Specifying 'H'; (b) specifying 'toggle'.

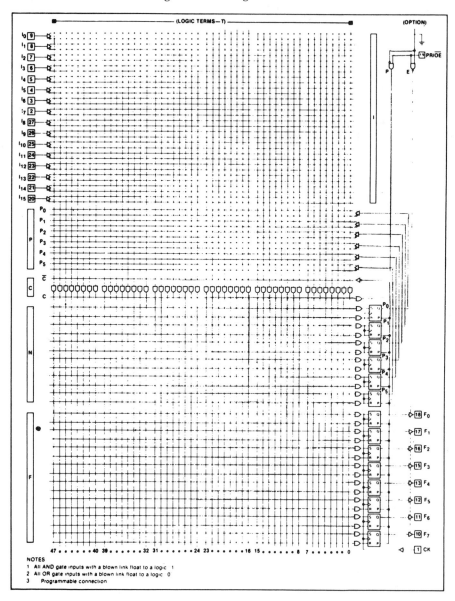

Fig. 5.20 PLS105 circuit diagram *(reproduced by permission of Philips semiconductors).*

$$Q2.J = /BIN * /Q3 * /Q2 * Q1 * /Q0$$
$$+ /BIN * Q3 * Q2 * Q1 * /Q0$$
$$Q2.K = /BIN * /Q3 * /Q2 * Q1 * /Q0$$
$$+ /BIN * Q3 * Q2 * Q1 * /Q0$$

$$Q1: = /BIN * /Q3 * /Q2 * Q0$$
$$+ /BIN * Q3 * Q2 * Q0$$
$$+ /BIN * Q1 * /Q0$$

$$Q0: = /BIN * /Q3 * /Q2 * /Q1$$
$$+ /BIN * /Q3 * Q2 * Q1$$
$$+ /BIN * Q3 * Q2 * /Q1$$
$$+ /BIN * Q3 * /Q2 * Q1$$

To complete the design we have to define the control terms needed to configure the flip-flops. Q3 and Q2 are J–Ks in both cases so we can blow the fuses to detach them from the flip-flop control line; Q1 and Q0 are J–Ks when BIN is HIGH and D-types when BIN is LOW so the flip-flop control equation is:

$$FC = BIN$$

All this information can be entered into a programming table as under:

I/P	Present state				Next state			
BIN	Q3	Q2	Q1	Q0	Q3	Q2	Q1	Q0
			F/F	type	.	.	A	A
H	–	–	–	–	–	–	–	0
H	–	–	–	H	–	–	0	–
H	–	–	H	H	–	0	–	–
H	–	H	H	H	0	–	–	–
L	L	H	L	L	0	–	–	–
L	H	L	L	L	0	–	–	–
L	L	L	H	L	–	0	–	–
L	H	H	H	L	–	0	–	–
L	L	L	–	H	–	–	H	–
L	H	H	–	H	–	–	H	–
L	–	–	H	L	–	–	H	–
L	L	L	L	–	–	–	–	H
L	L	H	H	–	–	–	–	H
L	H	H	L	–	–	–	–	H
L	H	L	H	–	–	–	–	H
FC H	–	–	–	–				

There are a number of points which come out of the above table. The 'F/F' term defines which fuses connecting the flip-flops to the 'FC' line are to be left intact; as with the OR array an 'A' denotes an intact fuse and '.' a blown fuse. The only valid next state entries for a D-type are 'H' and '–', otherwise there would be contention between the 'K' OR gate and the J-to-K inverter. Because we are not using the complement term or a preset or reset, these have been omitted for the sake of clarity. In a full table entry the complement term would have '–' in every line to remove it from the array; '0' (both fuses intact) is an illegal entry for the complement term as it could lead to oscillation. In order to make the other functions inactive we would leave all fuses intact, that is '0' in all entries in the programming table.

5.2.4 PLS availability

5.2.4.1 Overview

The PLSs listed in Table 8 form two major groups plus three individual devices which deserve separate consideration. The common architectural features of the two main groups have already been described, but again there are some individual characteristics which are not detailed in the table.

5.2.4.2 PLS105 family

The PLS105 was the first PLS to be introduced and the other members of the group are very similar. Technology improvements mean that the PLS105 itself is no longer outstanding either in speed or power economy. The PLUS105 vertical fuse device is much faster, while the CMOS PLSCE105 consumes half the power and is faster.

Other members of the group have enhanced architecture; the PLUS405, PSG506 and PLS30S16 all have more transition terms and registers, and variously contain features such as programmable clock polarity, programmable flip-flop type and dual register clocking. Dual clocking allows two separate state machines to be built in the same device.

Figure 5.20 shows the full circuit diagram of the basic PLS105.

5.2.4.3 PLS155 family exceptions

Sections 5.2.3.2 covered the PLS155 family in some detail; the two devices not covered there are the PLC415, which is a CMOS version of the PLUS405 with J–K flip-flops only, and the PLC42VA12. This device has an enhanced version of the PLS155 macrocell with an asynchronous clock option. This macrocell, shown in Figure 5.21, allows the PLC42VA12 to emulate the PLS179, PAL20RA10 and most PAL22V10 designs.

Figure 5.22 shows the circuit diagram of the PLS159.

5.2.4.4 GAL 6001

The input configuration of the GAL6001 may be selected as registered, latched or direct; the same options exist for the feedback from the I/O pins. Each section is treated as a separate block for the selection. By this means the device can capture data which may be transitory, for example from the data bus of a fast processor, provided a synchronising clock exists.

Each macrocell has a dedicated 'E-term' which provides an extra control

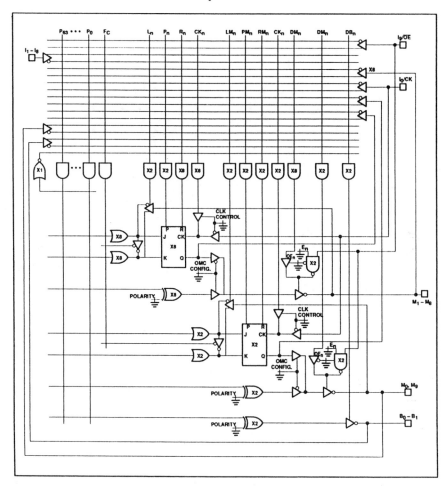

Fig. 5.21 PLC42VA12 macrocell *(reproduced by permission of Philips Semiconductors)*.

signal to the D-type register. In synchronous mode, which would usually be chosen for state machine implementation, the E-term acts as an enable and must be asserted before the flip-flop can be clocked. This has the same effect as a hold condition on the D-type and gives it the same power as an R–S flip-flop for defining state transitions. The E-term can also be routed to the clock input of the flip-flop, in which case the macrocell acts in an asynchronous mode. If the I/O pin is disabled, so that it becomes a dedicated input, the flip-flop feedback path is still available, allowing it to become a buried flip-flop.

5.2.4.5 XL78C800

A very different architecture is used by the XL78C800. Instead of an AND–OR configuration, it uses a folded NOR feedback scheme (Figure 5.23). The inputs and feedback signals are connected into a single NOR array;

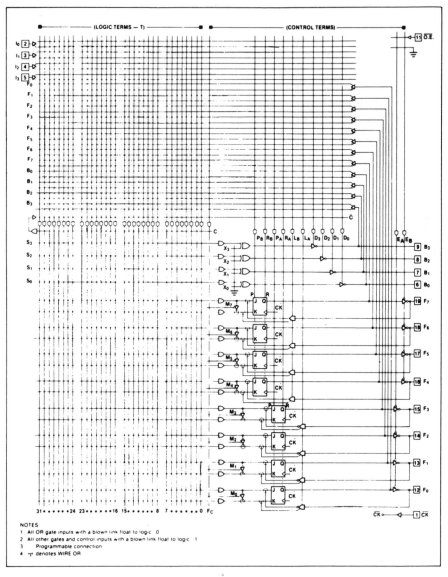

Fig. 5.22 PLS159 circuit diagram *(reproduced by permission of Philips Semiconductors)*.

part of the NOR array drives the output macrocells, and part drives gates which are fed back to the array itself. One possibility with this arrangement is multi-level logic, but it is equally suitable for implementing a standard state machine architecture. We saw in Section 2.1.3 that:

$$A * B * C = /(/A + /B + /C)$$

The NOR array can, therefore, behave as an AND array provided that the inputs are inverted. The macrocell, which is shown in Figure 5.24, has programmable invertors driving the registers so those NOR gates can be converted to OR. This makes the NOR–NOR directly compatible with AND–OR.

If we look at the macrocell in more detail, we see that a separate combinational output is provided. The output can be taken from this or from the output flip-flop. There are also two feedback paths, one from the flip-flop and one from either the combinational output signal or from the pin. This gives three ways of configuring the macrocell; registered output with registered feedback and a different combinational feedback, fedback combinational output with buried register and input with buried register. Each of these structures also have choice of output polarity and tri-state control.

Finally, although the flip-flops are dedicated J–K they may emulate a D-type by arranging that the J and K inputs are always complements, or a T-type by ensuring that they are always the same. Because the J and K terms have individual polarity control, these conditions can be implemented without using additional resources.

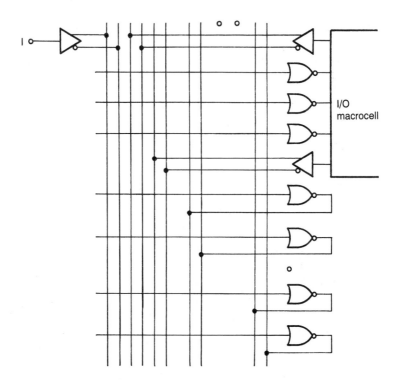

Fig. 5.23 Folded NOR feedback (XL78C800).

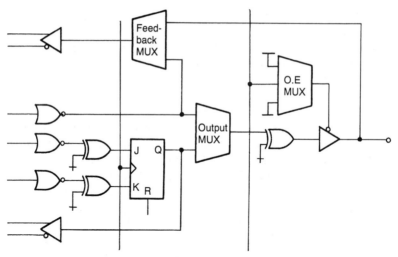

Fig. 5.24 XL78C800 macrocell.

5.2.4.6   7C361

A totally different approach to state machines is used in this device. Instead of encoding state information into, say, an 8-bit register each flip-flop in the state register represents a single state. The register outputs are fed back to a single array where they are combined with input conditions in a complex gate called a *condition decoder*. The state diagram has to be drawn in such a way that any state is entered by only one input condition, irrespective of the start state. This is shown in Figure 5.25, where the destination state, SD, can be reached from source states, S1, S2 and S3, when input condition, C, is TRUE. The entry condition for SD is thus:

$$(S1 + S2 + S3) * C$$

The condition decoder (Figure 5.26) has a NAND gate in place of the OR as this is a more compact structure in CMOS so, in practice, the entry condition is:

$$/(/S1 * /S2 * /S3) * C$$

C is a single product of the inputs. If the destination state is entered by different input conditions, a dummy state must be created which will be logically equivalent to the original destination. The same input conditions will trigger transitions from both the original and the dummy to any subsequent states, so the paths will then converge again in these states.

As well as controlled jumps, there is a direct path from each flip-flop macrocell to its neighbour. This is to allow for situations where a state is held until an exit criterion is met. The 'start' configuration for the macrocell, shown in Figure 5.27, allows the output to stay HIGH for only one clock cycle. If the adjacent macrocell is in the 'terminate' configuration (Figure 5.28), the next

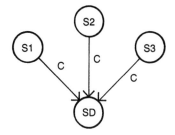

Fig. 5.25 7C361 state diagram – single destination.

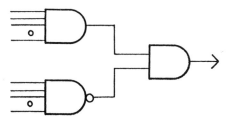

Fig. 5.26 7C361 condition decoder.

clock edge will pass the HIGH 'token' to it and this will be held until an exit condition becomes TRUE on the condition decode input. An exit condition will be of the form:

$$(C1 + C2 + C3) * SS$$

This is shown graphically in Figure 5.29 where S1, S2 and S3 are possible next states from the source state SS. Once again, in practice the OR equation has to be transformed to NAND to fit into the condition decoder.

There is a third possible configuration for the macrocell; that is 'toggle' configuration, shown in Figure 5.30. In this case the condition decoder or previous stage input feed either Q or /Q to the flip-flop input, causing it to respectively hold or toggle. This configuration can be used to build counters as the toggle condition for one stage of a counter is usually defined by a single product term.

The outputs are decoded from the state register by a standard OR array, but three different types of output are provided. A normal output provides for Moore machines and a bidirectional I/O gives extra input options. The third group of outputs gate the state sum term with an input signal, with the choice of AND, OR, exclusive-OR or no gating, plus output polarity. This gives the possibility of making Mealy machines.

This architecture is designed to give a higher speed option than standard PLSs. This is achieved by using only a single array in the feedback loop compared with the two (AND–OR) arrays in a normal PLS. A second factor is the size of the array. Rather than feed all macrocells back across the whole array, half are connected only to a local group of eight, a quarter to a larger group of sixteen and the last quarter only to all 32. This cuts the number of

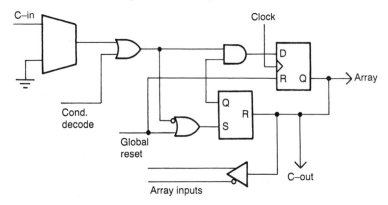

Fig. 5.27 7C361 'START' configuration.

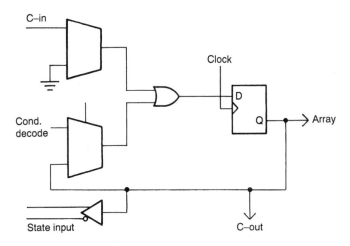

Fig. 5.28 7C361 'TERMINATE' configuration.

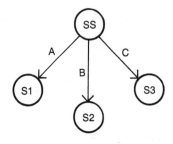

Fig. 5.29 7C361 state diagram – single source.

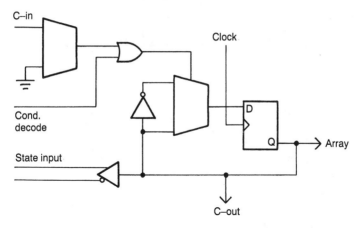

Fig. 5.30 7C361 'TOGGLE' configuration.

inputs per term to 56 from the 88 it would be if all feedback was global. The reduction in capacitance helps to keep the array delay to 8 ns which leads to a maximum operating frequency of 125 MHz.

A final unique feature of the 7C361 is the clock doubler. This doubles the clock frequency internally and allows state machine operation at twice the frequency of the system clock, provided that this is less than 62.5 MHz. Improved control is obtained in this way, even in systems using high speed processors, and for applications such as cache memory control.

5.3 EXAMPLES

5.1 Show how the programmable multiplexers in a GAL16V8 must be set for it to emulate:

(a) PAL12L6
(b) PAL16L8
(c) PAL16R4

5.2(a) Repeat the exercise of example 1 for an EP320.

(b) Highlight the differences between the GAL16V8 and EP320 in terms of architectural possibilities and potential applications.

5.3(a) Draw the state diagram for a three-floor lift. You will need a call signal for each floor, a sensor at each floor and a door-open detector; outputs needed are up and down for the lift motor, and open and close for the door motor.

(b) Write the equations for the lift controller as a registered PAL, and the truth table for a PLS.

Chapter 6
Programmable LSI

6.1 LSI PAL/FPLA STRUCTURES

6.1.1 Increasing PLD complexity

6.1.1.1 Scales of integration

It has been the natural tendency, in every integrated circuit technology, for the size and complexity of the circuits to increase. At first, while manufacturers are still learning to use a technology, simple circuits are made. An example is the simple gate circuit in TTL and CMOS. This level of integration is called SSI *(small scale integration)*. A comparable stage of development in programmable logic is the level of complexity described in Chapter 4.

The next stage is the ability to increase chip size economically, as the technology is better understood and controlled. Larger chips means more complex functions, such as multiplexers, counters and arithmetic circuits. Chapter 5 describes the equivalent level in PLDs; this is termed *medium scale integration* or MSI.

MSI circuits are still considered to be building blocks rather than complete functions. The LSI *(large scale integration)* stage is reached when complete functional blocks are made. Obvious examples from standard logic are the microprocessor, UART and speech synthesizer. This stage is harder to define in PLDs as many stand-alone functions can be built from the devices described earlier, particularly the sequencers.

The factors which make LSI stand out are a high I/O count (i.e. 40-pin package or larger), and an extensive logic capability (e.g. more than 150–200 product terms). Most of the devices described in this chapter meet these requirements.

6.1.1.2 Performance considerations

One feature that all the devices in Chapters 4 and 5 have in common is that the AND and OR matrixes, be they fixed or programmable, are a single entity equally accessible from all inputs and outputs. They are also all built in packages with no more than 28 pins. Allowing two pins for power supplies there are, at most, 52 MOS switches in every product term in the AND matrix, assuming all outputs are fed back to the AND matrix.

Propagation delay through a PLD may be split into four terms, input buffers, AND matrix, OR matrix and output macrocells/buffers. The only term which is affected by the I/O count is the AND matrix. Each MOS switch adds capacitance to the product term line, and therefore delay to the AND matrix. The geometry of PLDs is such that none of these delays dominates in the devices we have examined to date. However, as the number of MOS switches increases, the AND matrix delay starts to become the dominant factor.

Because LSI PLDs have a high I/O count, they must incorporate methods to limit the number of signals connected to the AND matrix. In the first part of this chapter we shall look at device architectures which still use a basic AND–OR logic structure, but have managed to increase their logic power without seriously compromising performance.

6.1.2 EP series

6.1.2.1 Overview

We have already met the two smallest members of this family in our discussions of enhanced PALs in Chapter 5. The EP320/330 are 'zero-power' generic PALs, while the EP610/630 are asynchronous PALs which also feature very low standby current. The two larger members of this family are the EP910 and the EP1810/1830.

The EP910 macrocell is identical in function to the EP610/630, but slower because each product term has 72 inputs compared with the 40 inputs of the EP610/630. The fastest of each type, made by the same process, are the EP610A-10 (i.e. 10 ns delay) and the EP910A-15. Otherwise the EP910 is just a stretched version of the EP610/630 with 12 dedicated inputs against 4, and 24 macrocells compared with 16.

The EP1810/1830 is the most complex member of the EP series of PLDs. With 48 configurable macrocells and 16 inputs it has over twice the logic power of the EP910, but uses a partitioned bus structure to reduce the number of inputs per product term from 128 to 88. Even so the fastest EP1810/1830 has a delay time of 20 ns.

6.1.2.2 EP1810/1830 bus structure

Figure 6.1 shows the overall block diagram of the EP1810/1830. The macrocells are grouped into four quadrants with 12 macrocells in each quadrant. Within each quadrant the feedback from every macrocell is available, together with all the inputs, for forming logic functions. Furthermore, one input is dedicated as the synchronous clock input for that quadrant, although that input still feeds the AND matrix and each macrocell can be configured in asynchronous mode.

Four macrocells in each quadrant are also connected to the product terms in the other quadrants, thus providing a logical connection across the whole device. Alternatively, this allows the macrocell to be buried within its own quadrant while its associated pin is used as a global input. The circuits of the

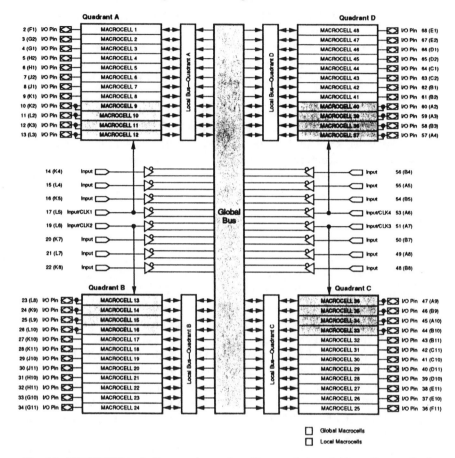

Fig. 6.1 EP1810/30 block diagram *(reproduced by permission of Altera Corporation)*.

local and global macrocell connections to the internal buses are shown in Figures 6.2 and 6.3.

This device can function as four more or less independent 12-macrocell PALs, or as a more complex PAL with limited global connection capability. The configurable macrocells give it additional versatility.

6.1.2.3 EP1810/1830 macrocell

The macrocell consists of three sections, a logic array, programmable flip-flop and tri-state I/O buffer. The logic array is based on a conventional AND–OR structure but is software configurable to implement a large number of combinational logic functions. It is claimed that each macrocell contains the equivalent logic power of 40 logic gates.

There are five possibilities for the output flip-flop. This may be a D-type, T-type, J–K, R–S or bypassed to give a purely combinational output. In

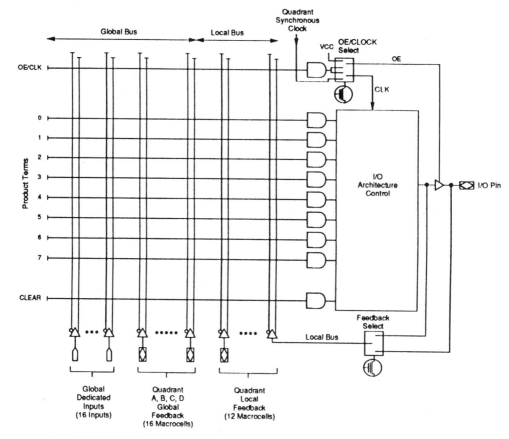

Fig. 6.2 EP1810/30 local macrocell connections *(reproduced by permission of Altera Corporation)*.

common with the other members of this family, one product term per macrocell can be allocated as an asynchronous clock (not the EP320/330) or an output enable. If it is used as a clock, the tri-state is set as permanently enabled or permanently disabled (i.e. input mode). These features are also evident from Figures 6.2 and 6.3.

6.1.3 ATV family

6.1.3.1 Overview

We also met the smallest ATV device when considering enhanced PALs. This was the ATV750; with 12 dedicated inputs and 10 double flip-flop macrocells, it has 84 inputs per product term and a fastest delay time of 20 ns. The ATV2500 is essentially a stretched version with 14 inputs and 24 macrocells. Since global connectivity is maintained, each product term has 172 inputs and the best delay time is degraded to 25 ns.

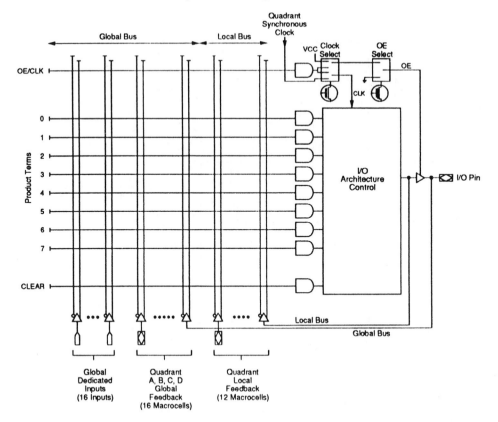

Fig. 6.3 EP1810/30 global macrocell connections *(reproduced by permission of Altera Corporation)*.

The product term allocation is different for the two devices. In the ATV750 there is a variable allocation from eight to sixteen terms split between the two OR gates, while the ATV2500 has twelve product terms per macrocells, shared between three OR gates. This arrangement allows both flip-flops to be buried with an active combinational I/O. As with the ATV750, some of the OR gates can also be combined to drive both flip-flops or the combinational output, if that mode is selected.

With 8 inputs and 52 macrocells, the ATV5000 is the most complex member of this family. As with the EP1810/1830, the macrocells are divided into quadrants but the logic arrangement is rather different.

6.1.3.2 ATV5000 bus structure

Each quadrant has 13 of the 52 I/O macrocells and six buried logic cells. The quadrants are connected by a universal bus which has only the feedback from the I/O cells as inputs. There is also a regional bus associated with each quadrant; it contains the eight direct inputs plus the feedback from all the flip-flops in that quadrant.

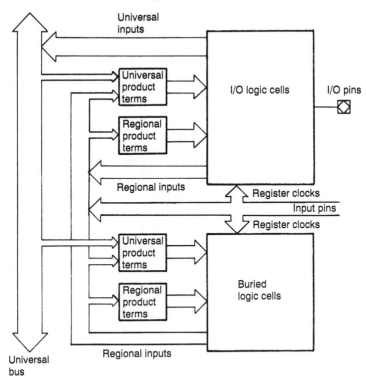

Fig. 6.4 ATV5000 block diagram.

Universal product terms include inputs from both the universal and regional buses; regional product terms include only inputs from the regional bus. The overall structure is outlined in Figure 6.4.

6.1.4 MACH family

6.1.4.1 Overview

MACH stands for Macro Array CMOS high-density and is the first example we have come across in this chapter of an LSI family with no lower density precursors. As with the two previous examples, the logic is partitioned internally into blocks but the routeing of the logic signals is achieved in a different manner. To understand the full potential of the MACH devices we must first examine two new concepts in PLD architecture.

6.1.4.2 The switch matrix

The switch matrix may be likened to the nerve centre of a MACH device. It performs no logic function in the sense that there are no gates or flip-flops, but it controls the routeing of every signal to the inputs of the logic blocks. The

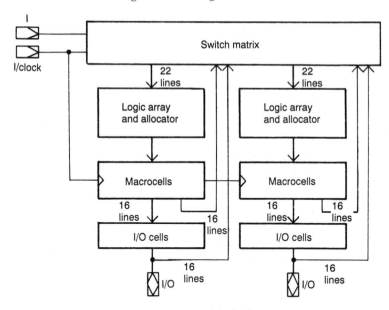

Fig. 6.5 MACH110 block diagram.

simplest MACH device is the MACH110 and by referring to its block diagram, in Figure 6.5, we can illustrate the function of the switch matrix.

The MACH110 has 32 macrocells, grouped into two blocks of logic called PAL blocks; there are also six dedicated inputs, two of which double as clocks. Since feedback is taken from both the macrocell and the I/O cell, there are 70 logic signals available internally for combining into logic functions. Just 22 of these signals are routed by the switch matrix into each of the two PAL blocks.

This has the advantage that each product term has only 44 inputs; a fastest propagation of 15 ns is achieved as a result. This delay is independent of the path through the device since all signals are routed by the same path.

6.1.4.3 Logic allocator

The second novel concept used in the MACH family is the logic allocator. Each PAL block has just 64 product terms to share between its 16 output macrocells. The logic allocator routes the product terms, in groups of four, to the macrocells to a maximum of 12 per macrocell. Its function is illustrated in Figure 6.6.

This parsimonious allocation can, by inference, make some of the macrocells unusable as outputs in spite of the double feedback allowing input pins to bury their macrocell. However, with just six direct inputs, it is likely that not all 32 macrocells will be needed and the allocator will make the most efficient use of the available logic power.

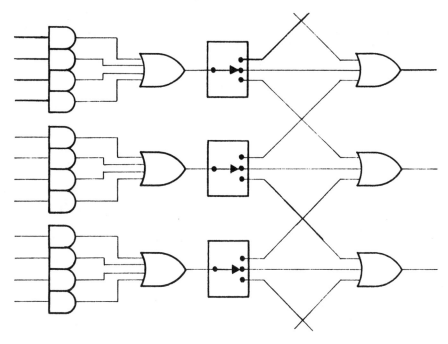

Fig. 6.6 MACH logic allocator.

6.1.4.4 Family description

The output macrocells themselves are fairly basic PAL-type components. The flip-flop can be a D-type, T-type or bypassed, although the more advanced MACH-2 devices allow a latch configuration as well. The MACH-2 devices also have a slightly different PAL block, in which half of the macrocells are buried and half connected as I/O cells. The larger members of the family also have 26 inputs to each of the PAL blocks, which gives them a slightly more powerful logic capability.

In all, six members of the MACH family have been defined to date. Each has six or eight dedicated inputs; the MACH-1 devices have 32, 48 and 64 output macrocells respectively. The same pattern runs through the MACH-2 devices, but these have as many buried macrocells as outputs and, therefore, approximately twice the logic power of the equivalent MACH-1 device.

6.1.5 MAX family

6.1.5.1 General description

The MAX family (MAX stands for Multiple Array Matrix) has its antecedents in the EP-series but is more similar to the MACH family. The global bus of the EP1810 is replaced by a switch matrix, called a programmable

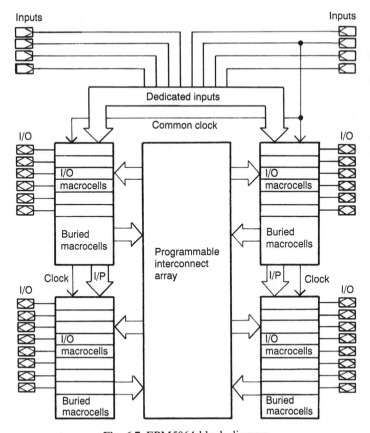

Fig. 6.7 EPM5064 block diagram.

interconnect array. MAX devices have a macrocell with product term derived clock, preset, clear and output enable, but only four logic terms (one drives one side of an exclusive-OR gate) compared with the eight terms of the EP-series. The 'spare' terms are placed in an expander for use by any macrocell in the logic array block (LAB), as the 'PAL block' is called, thereby doing away with the need for a logic allocator function, which it effectively replaces.

We can look at the component parts of a typical MAX device, the EPM5064 also known as the 7C343, with a block diagram shown in Figure 6.7.

6.1.5.2 *Programmable interconnect array*

The PIA is the switch matrix which routes all the internal signals to the logic array blocks. The feedback signals from all 64 macrocells are connected to the input side of the PIA together with the inputs from the 28 I/O pins, making a total of 92 input lines. As in the MACH family, some of these may be used as inputs to the LABs.

Unlike the MACH devices, all the direct inputs and the local macrocell feedback signals are available as inputs to each macrocell. This is a mixed blessing. On one hand it ensures that all local signals and direct inputs are always available without using up interconnect resources to the other LABs. The disadvantage is that the delay times depend on the signal path.

For example, a direct input signal will take 25 ns to arrive at the register input (maximum specified time in a standard EPM5064), a signal from a macrocell in the same LAB 18 ns, from a macrocell in another LAB 38 ns and from any I/O pin 45 ns. This can clearly lead to timing problems, particularly if the system is being run at the fastest specified clock speed of 33.3 MHz, that is a clock period of 30 ns.

6.1.5.3 Logic array block

Figure 6.8 shows the circuit diagram of one macrocell and part of the expander array. Each macrocell has four control terms (no output enable for buried macrocells) and four logic terms. The logic terms are split three-to-one to either side of an exclusive-OR gate. As we have seen before, this allows the

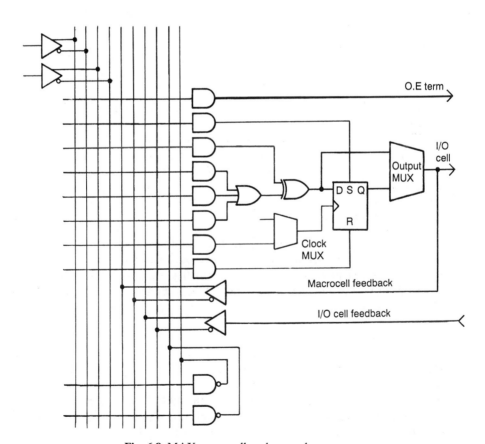

Fig. 6.8 MAX macrocell and expander array.

flip-flop to behave as a J–K type, or the exclusive-OR can be used to program the polarity of the logic signal into the D-type flip-flop.

The expander has 32 product terms in each LAB and is equivalent to a 32 gate FPLA. Conventionally, an FPLA defines logic by AND–OR terms such as:

$$X = A * B + C * D + E \ldots \ldots$$

By performing a DeMorgan transformation this can be rewritten as:

$$/X = /(A * B) * /(C * D) * /(E \ldots \ldots$$

This is just the logic arrangement of the expander block, with an optional inversion at the exclusive-OR gate. Once again though care must be taken with the timing as the expander can add as much as 20 ns to the signal delay. Thus, a direct input not using the expander will be available at the flip-flop input in 25 ns, as above, but an I/O input using the expander will take 65 ns.

6.1.5.4 MAX family description

The numbering system for the 5000-series family is quite logical as the part numbers take the form EPM5*nnn* where *nnn* is the number of macrocells in the device. Some manufacturers do use a different scheme. The smallest MAX device is the EPM5016 with eight inputs, eight I/O pins and a single LAB of 16 macrocells. Because it has only one LAB there is no need for a PIA, all internal signals being available to all macrocells.

The most complex device in the family has eight inputs and 12 LABs, four with eight I/O pins and eight with four I/O pins. The EPM5192 thus has 64 potential outputs and at least 128 buried macrocells.

All MAX devices have 8 inputs except the EPM5130 which is a 20-input version of the EPM5128 and therefore defies the logical numbering sequence of the family.

6.1.5.5 EPM7000-series

A second generation MAX family has been announced as the EPM7000-series. Rather than being numbered by the macrocell count they are numbered by a claimed gate equivalent, from 1500 to 20 000 gates. They will be made by a faster process, and enhancements to the LAB and PIA should remove some of the timing skews.

6.1.6 Plus logic

6.1.6.1 Plus architecture

Plus logic arrays have much in common with MACH and MAX families. They have three basic sections; the I/O block provides an interface to the outside world. It is driven by the functional blocks, and itself feeds a UIM (universal interconnect matrix) which provides the major interconnections

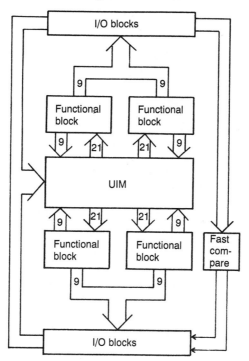

Fig. 6.9 Plus logic architecture.

inside the array. Unlike MACH and MAX, the UIM also provides a single-level logic resource.

An example of this architecture is shown in Figure 6.9, which pictures the simplest Plus logic device – the FPGA2010. The capabilities of this family may be seen by examining the three separate features individually.

6.1.6.2 I/O block

Some of the I/O block pins are dedicated inputs or outputs, but most are true I/Os. An I/O pin schematic is shown in Figure 6.10; dedicated inputs and outputs contain the appropriate half of the I/O logic.

Inputs can be direct, clocked or latched, although the clock/enable signal must be one of the direct clock inputs called FastClocks. These buffered clock inputs, on shared output pins, offer a lower delay than product term clocks when a common clock is required for functional block registers.

Outputs, including those making use of unused FastClock pins, have an optional tri-state capability. The enable for this can be derived from a product term, which may be gated with the FastOE signal. This has similar properties to the FastClocks.

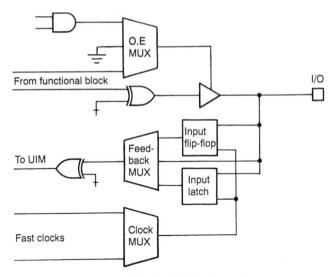

Fig. 6.10 Plus logic I/O schematic.

6.1.6.3   The UIM

Every I/O pin and functional block output acts as input to the UIM, which is, effectively, a large AND array. Twenty-one outputs from the UIM feed a second AND array associated with each functional block; outputs from the second array provide the input signals to each individual macro. Figure 6.11 shows this structure in schematic form.

A typical Plus FPGA, the 2020, has eight function blocks, each with nine macros. The number of inputs depends on the package, but 72 I/O blocks are possible so the UIM will have 144 inputs and 168 outputs in this device, quite a formidable logic capability without considering the functional blocks and macros.

6.1.6.4   Functional blocks

Each functional block has a 42-input 57-output AND array, which drives the individual macros. The schematic of a single macro is shown in Figure 6.12. At the heart of the macro is a logic expander which can combine its two inputs in any standard Boolean function. Alongside this is a 'carry function' which takes the same two inputs, together with a carry-in from an adjacent macro, and produces a carry-out which is passed to the next macro. Thus, if the expander is programmed with the exclusive-OR function, the macro becomes a full adder cell.

The two inputs to the logic expander are formed by OR-ing some of the outputs from the functional block AND array. Four 'private' and four shared outputs form one input, while the second is derived from one private term and

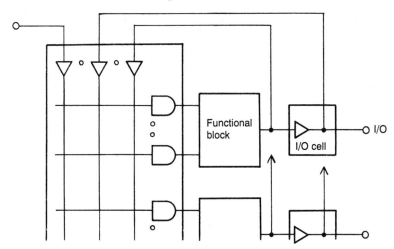

Fig. 6.11 Plus logic UIM.

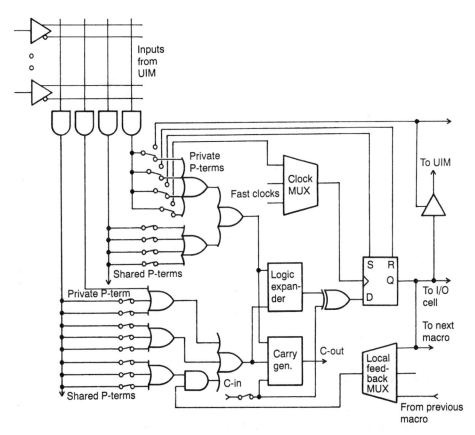

Fig. 6.12 Plus logic functional block.

eight shared terms. If any of the four private terms in the first input are not used as logic they may be used as control terms for the macro flip-flop.

Each macro contains a D-type flip-flop which may be clocked from an unused logic term or by one of the FastClocks. Other unused terms may be used as set, reset or as an output enable. Feedback to the lower logic expander input may be selected from the flip-flop of the same macro or from the adjacent macro. This could allow a shift register to be built without using the UIM for connecting stages.

6.1.6.5 *Additional logic capability*

A number of the inputs, depending on which package is used, may be connected to a dedicated comparison circuit, shown in Figure 6.13. The input is compared with a bit of data which has previously been latched into the comparison circuit. If all bits compare equal, an output is set for internal or external use.

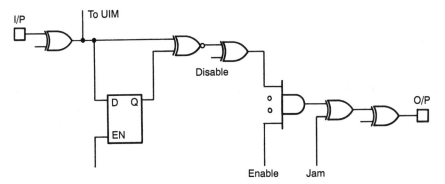

Fig. 6.13 Plus logic comparator.

Two members of the Plus logic family also contain extra fixed resources. The FPSL5110 contains 1152 RAM bits which may be configured as 32×36, 64×18 or 128×9. Logic with embedded RAM makes it possible to construct very fast structures such as dual port RAM or FIFOs with surrounding logic functions.

The second device is the FPSL5210 which has a 128-state state machine alongside the logic array. High speed is again the benefit as this dedicated structure will function at a higher speed than a state machine relying on feedback via the UIM.

6.1.7 MAPL family

6.1.7.1 *Overall description*

The MAPL family uses a different technique for maintaining a reasonable speed performance in a complex device. Increase in propagation delay through programmable arrays is caused by increasing the number of programmable connections to each term in the array. This is because of the extra capacitance which has to be charged and discharged when the product or sum term changes its logic level. This effect can be reduced by increasing the standing current in the term. A higher current will increase the rate at which charge is added or removed from the capacitance.

Increasing the current does cause other problems. There is a limit to the power dissipation possible in a given package without causing the chip inside to overheat. High power dissipation also makes added demands on the system power supply and heat removal from the box containing the final system.

In MAPL devices these problems are resolved by only powering an eighth of the total array at any one time. Figure 6.14 shows the block diagram of the MAPL144.

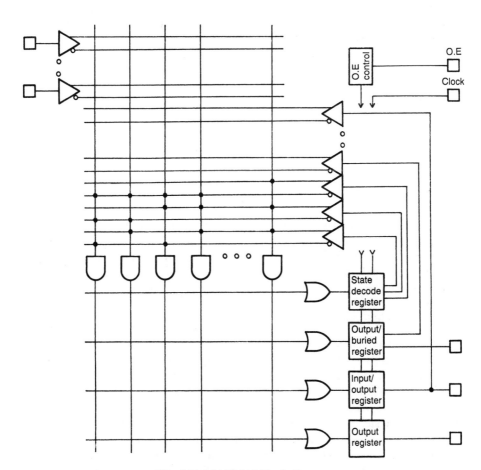

Fig. 6.14 MAPL144 block diagram.

6.1.7.2   *MAPL arrays*

The MAPL144 is just a large FPLS. In all it has 128 transition terms and three types of macrocell. The macrocells, in Figure 6.15, give the option of either J–K or D–E type flip-flops, resettable by a product term, and six output enable choices. The output, where available, can be always enabled, always disabled, controlled by the OE input or controlled by one of three product terms. There are two types of logic macrocell, an I/O cell whose pin can be used as an input or a fed back output, and an output/buried cell whose feedback is taken before the output enable buffer.

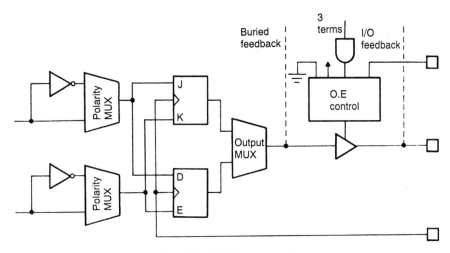

Fig. 6.15 MAPL macrocell.

The third macrocell is the global state cell. There are just three of these, and their outputs are decoded internally to define which 'page' of the logic array is switched on for the next group of transition terms to be active. Clearly this implies that all the terms associated with one state must be located in the same page of the array.

6.1.7.3   *Family summary*

There are two members of the basic MAPL-1 family. Both have nine inputs, twelve I/O cells and four output cells with eight pages of sixteen AND–OR terms. The MAPL128 has eight buried macrocells, but in the MAPL144 these are also available as outputs. Both devices have speed versions capable of running at 45 MHz with only 140 mA supply current.

One drawback of this family is that only registered outputs are available. This has led to the definition of a MAPL-2 family which has an eight-output GAL block added to the basic FPLS structure.

MAPL is an acronym of Multiple Array Programmable Logic.

6.1.8 Programmable macro logic

6.1.8.1 PML concept

Programmable macro logic (PML) is a freer structure than the conventional PAL/PLA architectures described so far. It is based on a core of foldback NAND terms similar to the expander array in the MAX family, although PML was defined at an earlier time (1985) than MAX. As we saw in the MAX description, NAND–NAND is logically equivalent to AND–OR, so PML implements logic functions in the same way as PALs and PLAs. Because every term is fed back internally PML has the potential for more efficient usage of logic terms because multi-level logic can be defined without using up other resources.

The fundamental architecture of PML is shown in Figure 6.16. The logic core is surrounded by input, output and internal function macros. The present range of PML devices includes no functions that cannot already be found in PLDs, for example, exclusive-OR gates and flip-flops, but it is not difficult to see the potential of including more extensive arithmetic functions, RAM modules and so on.

This is a significant step forward for PLDs were originally seen as replacement for the discrete glue logic surrounding LSI functions. With PML, the possibility exists for integrating the glue logic with the LSI functions themselves.

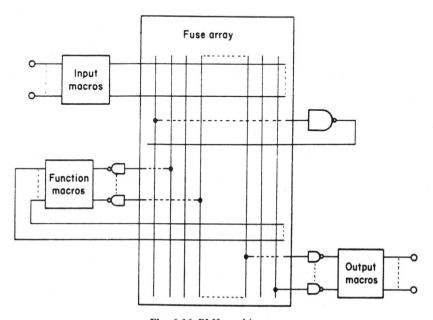

Fig. 6.16 PML architecture.

6.1.8.2   PML family description

Four devices have been defined in PML to date. Two of these, the PLHS501 and PLHS601 are solely combinational, the PLHS501 with 24 inputs, eight I/Os and 16 outputs has 72 foldback NAND terms; the PLHS601 has 78 NAND terms as well as eight exclusive-OR functional macros, 28 inputs, 12 I/Os and 12 outputs. The PLHS502 has similar I/O resources as the PLHS501, but includes eight D-type and eight R–S flip-flops as internal functions.

These three devices are unusual among LSI PLDs, being made in bipolar technology. They operate at moderate speeds, the flip-flops can be clocked at 40 MHz and propagation delay from input to output with a double array pass is about 40 ns, but supply currents are very high – 400 mA for the PLHS502.

The fourth PML device, PML2552, is a CMOS device. Its supply current is more reasonable at 120 mA and the flip-flops can be clocked at 50 MHz, but I/O delay with a double array pass is 60 ns. The PML2552 logic resources include 29 inputs, 16 of them registered, 24 I/Os, with 16 registered again, 20 internal J–K flip-flops and 96 foldback NAND terms.

A special feature of the PML2552 is that it incorporates scan path testing, similar to a diagnostic PROM. In diagnostic mode the internal and output registers are configured as a shift register which allows internal data to be read in serially, an application run, then the result read out. This feature makes the PML2552 easier to test than other PML devices, and other LSI PLDs which do not have scan path test capability.

6.1.9   Performance comparison of LSI PLDs

6.1.9.1   Summary table

While the varied features make tabulation difficult, for comparison purposes, Table 9 lists many of the properties and performance data of the LSI PLDs discussed above.

6.1.9.2   Benchmark design

One way to compare different products is to use a *benchmark test*. In this the same design is made in different device types and the result compared in terms of resource usage within each device and device performance. A third factor is device price, but we will not be considering that here as prices change frequently and may often be consolidated with other devices in a package deal, so a stand-alone price may not be the best that a user can obtain from a particular manufacturer.

The choice of a benchmark design is not straightforward. When this test is used by a manufacturer to compare his product against the competition, he will naturally choose a circuit which shows his device up in a good light. Any design may favour one family unwittingly, showing a bias which may be misleading in terms of some other design.

One approach is to use a series of circuits, each of which has a different

Part	Inputs		Macrocells			AND	Performance	
Number	Dir.	Reg.	I/O(+ bur.)	O/P	Bur.	Terms	$f._{max}$	I_{cc}
EP Series								
EP320/330	10	—	8	—	—	8/PAL cell	100	75
EP610/630	4	—	16	—	—	8/PAL cell	100	130
EP910	12	—	24	—	—	8/PAL cell	83	150
EP1810	16	—	48(+ 16)	—	—	8/PAL cell	50	200
5AC Devices								
5AC312	—	10	12	—	—	8/PAL cell	100	33
5AC324	—	12	24	—	—	8/PAL cell	175	33
ATV Family								
ATV750	12	—	10(+ 10)	—	10	6/f-f(avge.)	55	15
ATV2500	14	—	24(+ 24)	—	24	6/f-f(avge.)	40	120
ATV5000	8	—	52(+ 52)	—	76	tba	50	50
MACH Family								
MACH110	6	—	32(+ 32)	—	—	4/PAL cell(av.)	66	150
MACH120	8	—	48(+ 48)	—	—	4/PAL cell(av.)	66	180
MACH130	6	—	64(+ 64)	—	—	4/PAL cell(av.)	66	180
MACH210	6	—	32(+ 32)	—	32	4/PAL cell(av.)	66	180
MACH220	8	—	48(+ 48)	—	48	4/PAL cell(av.)	tba	tba
MACH230	6	—	64(+ 64)	—	64	4/PAL cell(av.)	66	300
MAX Family								
EPM5016	8	—	8(+ 8)	—	8	4/cell + 32 exp.	100	115
EPM5032/7C344	8	—	16(+ 16)	—	16	4/cell + 64 exp.	77	155
EPM5064/7C343	8	—	28(+ 28)	—	36	4/cell + 128 exp.	50	135
EPM5128/7C342	8	—	52(+ 52)	—	76	4/cell + 256 exp.	50	250
EPM5130	20	—	64(+ 64)	—	64	4/cell + 256 exp.	50	275
EPM5192/7C341	8	—	64(+ 64)	—	128	4/cell + 384 exp.	50	380
MAPL Family								
MAPL128	9	—	12	4	8	128 PLA terms	45	140
MAPL144	9	—	12	12	8	128 PLA terms	45	140
PLUS Family								
PLUS2010	n/a	n/a	36	n/a	n/a	5/cell + 12 com	40	n/a
PLUS2020	—	3	72	—	—	5/cell + 12 com	40	260
PLUS2040	n/a	n/a	144	n/a	n/a	5/cell + 12 com	40	n/a
FPSL5110	as PLUS2010 with 1152 configurable register bits						40	n/a
FPSL5210	as PLUS2010 with 128 state finite state machine						40	n/a
PML Devices								
PLHS501	24	—	8(comb.)	16(comb.)		72 PML + 32 o/p	30	295
PLHS502	24	—	8(comb.)	16(comb.)	16	88 PML + 24 o/p	33	400
PLHS601	28	—	12(comb.)	12(comb.)		94 PML + 24 o/p	45	340
PML2552	13	16	24	—	20	136 PML + 24 o/p	50	120

Table 9   LSI PLDs.

emphasis. For example, the logic requirements of state machines and arithmetic circuits are quite different. Similarly, random logic and counters need a different mix of components and architecture. Comparisons may be made by using an example of each type of circuit and seeing how many times it can be fitted into the target devices. Alternatively, a test circuit containing each variety of circuit configuration may be designed and an estimate made of the proportion of resources used by each target device.

6.1.9.3 Test circuit

The test circuit should meet certain criteria. Firstly, it should be easy to understand, then it should not be too complex, otherwise some of the families may not cope at all. Lastly, it should not be too simple either, or it will not stretch the resources sufficiently for any differences to show. A possible circuit is shown in Figure 6.17, drawn in terms of 74TTL family components. It includes a 16-bit multiplexer, two 4-bit counters, two 4-bit adders, a dual J–K flip-flop and some gates. Its function is probably not very useful in itself, but it is easily understood and should provide a reasonable challenge for LSI PLDs.

The circuit will count how many HIGHs there are in the 16-bit input word, but it is also capable of being chained with similar circuits to provide a grand total for up to 256 bits of input.

Input 'dvd' sets one flip-flop which allows one of the counters to start counting. The count output selects one of the 16 input lines via the multiplexer; if this is HIGH the second counter increments. Once the first counter reaches 15 the second flip-flop is reset and holds the count, which is added to the 8-bit carry input in the two 4-bit adders. The output from the second flip-flop is called 'ovd' and signifies that the output is valid.

Potential problems with PLD implementation are the multiplexer and the adder. PLDs with no PLA terms may have to use internal feedback and two or more passes through a PAL array to cope with the 16 product terms in the multiplexer. The adder poses a similar problem, but is also not a simple design. A 'wide' adder, in AND–OR configuration, is very inefficient in logic usage, so a 'serial' design is probably better. In this case, each pair of bits is added, starting with the LSBs, and the carry used as an input for the next pair of bits in order. This method is slow, the total delay being eight times the delay through one stage, but the result will be a good indication of the overall speed of the target device.

The percentage of product terms and internal feedbacks used will show how much 'logic power' the device contains. With 27 inputs and nine outputs, altogether, the circuit should fit into a 40/44-pin member of the respective PLD family.

6.1.9.4 Benchmark test result

From the information available, this design will only fit conveniently into four of the device families described so far. For comparison purposes we have studied the MACH210, EPM5064, ATV2500 and PML2552 as potential

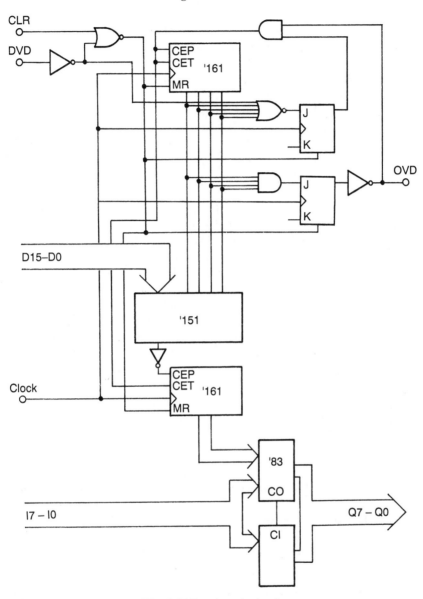

Fig. 6.17 Benchmark circuit.

targets. Except for the PML2552 they are all 40/44 pin devices; the PML2552 is the only member of the PML family with sufficient resource, but is a 68-lead device.

The reasons for 'failure' among the other families are as follows. The EP-series was not considered as the EP910 has insufficient resources while the EP1810, with 68 leads, is made by the same manufacturer as the EPM5064. The 5AC324 has no buried macrocells and the MAPL144 has too few I/Os.

Part	I/O		Product terms			Macrocells			% Used	
number	Tot	Used	Tot	Used	Free	Tot	Used	Free	P-ts	Cells
MACH210	38	36	256	84	136	64	30	34	47	47
EPM5064	36	36	320	62	192	64	27	37	40	43
ATV2500	38	36	288	86	144	24	18	6	50	75
PML2552	54	36	132	69	63	—	—	—	53	—

Table 10   Benchmark test results.

The results are summarised in Table 10 and, as might be expected, the number of product terms used is similar for devices using the same type of flip-flop. The MACH210 and ATV2500 are broadly similar as both use D-types. The difference comes from the way in which the 16-input multiplexer is built. The logic allocator in the MACH allows 16 product terms in one macrocell, but the ATV2500 must use two macrocells to build it. The product term figures for the PML2552 count 'J' and 'K' as one term, and the exclusive-OR terms have been ignored in the EPM5064 as they serve to configure the flip-flop as T-type. Extra terms are required by the PML2552 to generate and feed back the carry terms in the adder. Usable terms in the EPM5064 depend on how the macrocells are allocated among the logic blocks; this will affect the number of expander terms which are still available.

It appears that the EPM5064 is the least filled device. The EPM5064 has to use expander terms in both the multiplexer and the adder sections, so it will also be the slowest solution. It will always be worthwhile investigating all the possibilities in order to find the optimum solution to any logic problem.

6.2   PROGRAMMABLE GATE ARRAYS

6.2.1   Gate array architecture

6.2.1.1   Logic cell description

Programmable gate arrays are a relatively new aspect of programmable logic. Gate arrays themselves have been in existence for over a decade but, until recently, they were only available as masked devices. As such, they were subject to long lead times, high development charges and medium to high quantity restrictions. These are all factors which PLDs are designed to overcome, so the arrival of field programmable versions is sure to create new markets for this high density logic architecture.

The heart of any logic device is the 'logic engine'. In conventional PLDs this is the AND-OR matrix and the logic macrocell. Gate arrays, on the other hand, use low density logic functions embedded in an extensive interconnect matrix, the so-called 'sea of gates'. In masked arrays each logic cell, in its undefined state, is a collection of p-channel and n-channel transistors. These are connected, by the first metallisation layer, to form whatever function is called up by the particular design being implemented. Each cell might be configured as a complex gate, latch, flip-flop or similar SSI function.

This approach cannot be used for PGAs as all the metallisation layers must be defined before the device is encapsulated in its final package. Instead, each cell must consist of a universal logic element which is capable of being configured into as many basic functions as possible. This basic logic cell is commonly known as a 'configurable logic block' or, more simply, a logic module or logic cell.

6.2.1.2   *Gate array interconnect*

Interconnecting the logic cells in masked arrays is relatively straightforward. The metallisation, which is used for defining the logic cell function, is also used for making connections between the logic cells. This does not usually give enough interconnect capacity, nor allow for crossovers across tracks, so a second layer of metal is deposited over the first with an intermediate layer of silicon dioxide providing insulation. Holes, called *vias*, are etched in the silicon dioxide to provide connections between the two layers.

Once again, this technique is not possible in PGAs. Provision must be made for the connection of any one logic cell to any one of several others, although not necessarily directly to every other cell in the array, so that the design is not limited unduly by interconnection restrictions. The 'horizontal' and 'vertical' spaces (strictly the x and y directions) between the cells are filled with aluminium stripes which extend whole or part way across the array, and have programmable interconnections where they cross. Two techniques are commonly used for the programmable connections; one is the anti-fuse described in Chapter 3, the other is described in the following section.

6.2.1.3   *RAM-based interconnect*

Random access memory (RAM) is not a structure we have discussed yet, in connection with PLDs. It is based on an inverter loop configuration, Figure 6.18, which has two stable states. If the input of one inverter is HIGH the other inverter will have a LOW input which will maintain the state of the first. When the transmission gate is open, a signal on the input line can force the pair into the opposite state.

The outputs can be connected to transmission gates or multiplexers; transmission gates can form programmable interconnections, while multiplexers are used to define logic paths in macrocells. The sense of each RAM cell will, therefore, define the architecture and connection pattern of the gate array.

There are two drawbacks to this method. One is architectural in that a separate interconnection network has to be provided to program all the RAM cells, which need four components at each crossing of two tracks. Because the same technique must also be used for defining the function of each logic cell,

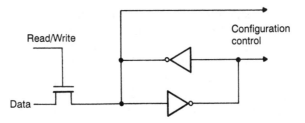

Fig. 6.18 Simplified RAM cell.

by means of configuration connections, this method is likely to be inefficient in terms of the chip area used for a given logic complexity.

The second drawback could be turned into an advantage in certain circumstances. When power is removed from a RAM cell it loses the memory of its last state; thus, when power is restored, the configuration will be different and may be random unless steps are taken in the cell design to ensure that it always goes to the same state when power is first applied. When the equipment in which the array is installed is switched off the logic information in the array will be lost, so the RAM must be reloaded every time the equipment is switched on.

This procedure is simplified by provision of serial PROMs which can be programmed with the array data, permanently connected to the data inputs of the array and automatically download their data on switch on.

The ability to define the logic content of the array by loading the array RAM with data makes it possible to change the function of a device as easily as changing the program in a computer. Indeed, it means that a computer could configure its own hardware to optimise it for a given application.

6.2.2 Logic cell array – 2000 series

6.2.2.1 *Internal structure*

The logic cell array has three principal types of internal component. There are I/O cells for interfacing to other external components, logic cells (called configurable logic blocks) and the interconnection array. The CLBs are arranged in a square matrix so that the LCA2064 has 64 in an 8 × 8 arrangement, and the LCA2018 has 100 set out as 10 × 10. The spaces between the cell rows and columns contain the interconnect, while the I/O cells are arranged around the perimeter. Figure 6.19 shows this structure.

6.2.2.2 *Configurable logic block*

The core of any PGA is the logic cell. The CLB circuit is shown in Figure 6.20 and contains a combinational block, a storage element and multiplexes for routeing signals inside the CLB and to and from the interconnection matrix.

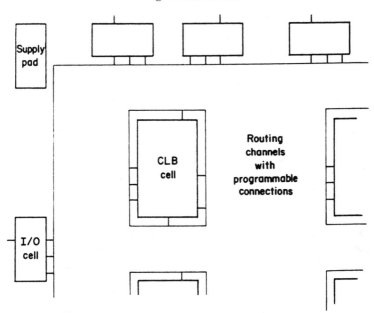

Fig. 6.19 LCA block diagram (one corner).

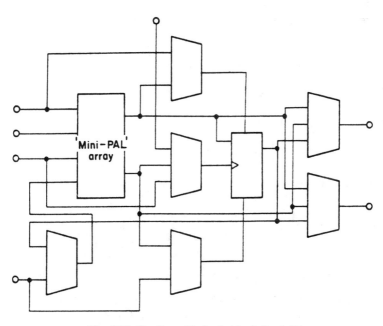

Fig. 6.20 Configurable logic block for LCA.

The combinational block is like a **PROM** inasmuch as the logic function is defined by a look-up table inside it. It does not have the full versatility of a **PROM** as the outputs are restricted to identical four-input functions or distinct three-variable functions. There are four inputs to the combinational block from the interconnect, but one of these (D) is shared with the feedback from the storage element. The two outputs can be taken directly to the interconnect or to the storage element.

The storage element can be either a D-type flip-flop or a D-latch. The D input must be one output (F) from the combinational block, but the clock/enable can be an output (G), a common input (C) with the combinational block or a direct input from the adjacent interconnect line. Set and reset to the storage element are derived from either common inputs with the combinational block or its outputs; the storage element output can be fed back to the combinational block or used as an output to the interconnect.

6.2.2.3 *I/O cells*

The I/O cell schematic, Figure 6.21, holds no great mystery. Signals are routed into the array via an optional flip-flop with a common clock. The outputs from the array pass through a tri-state buffer, which may be controlled, always open or disabled, in which case the pin will be a dedicated input.

6.2.2.4 *Interconnect array*

The interconnect array routes signals in three different distance scales; these are local connection, general purpose interconnect and long lines.

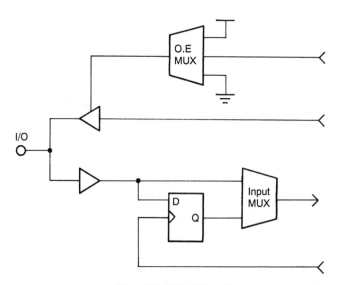

Fig. 6.21 LCA I/O cell.

Local connections provide a high speed signal path between adjacent CLBs, but only between restricted sets of inputs and outputs. For example, the 'X' output may be connected to the 'C' or 'D' input of the cell above it, or to the 'A' or 'B' input of the cell below. The 'Y' output can be connected to the 'B' input of the CLB to its right. Direct connection is also possible between I/O cells and CLBs at the edge of the matrix.

General purpose interconnect uses the 'channels' between rows and columns of CLBs. Four horizontal lines between each row and five vertical lines between each column meet in a switch matrix at every crossing point. A signal entering from the 'north' of the switch matrix can be routed to 'east', 'south' or 'west', with certain restrictions. The signals can be switched into or out of adjacent CLBs by means of transistor switches on the interconnect. Buffers are provided at intervals on the general purpose interconnect to compensate for the higher capacitance of the interconnect/switch matrix combination.

The long lines use the same channels as the general purpose interconnect, but bypass the switch matrices. Each long line travels the whole length or width of the array, and they are primarily intended for distributing common signals among the CLBs. In particular, one line in each channel can be directly connected to the clock input of each CLB and, therefore, provides a low skew, well synchronised global clock signal.

6.2.2.5 Family members

There are just two devices in the LCA2000 family. The LCA2064 has 64 CLBs and 58 I/O pins, while the LCA2018 has 100 CLBs and 74 I/O pins. They are available in speed versions of up to 70 MHz but the power consumption depends on the number of cells used and the operating frequency. The standby currents are 10 and 15 mA respectively, but a fully utilised device running at maximum frequency could require a supply current of over 500 mA.

The number of RAM bits in each device are 12038 (LCA2064) and 17878 (LCA2018); the serial PROM type 1736 contains 36288 bits and interfaces directly to either LCA. Multiple LCA designs can use a single serial PROM as there is a 'daisy-chain' output which allows data to pass through one LCA to load a second (or third, etc.) directly.

6.2.3 LCA3000 series

6.2.3.1 Comparison with 2000 series

The LCA3000 series is a second generation extended version of the LCA2000 family. The 'floor plan' is very similar to the LCA2000 series, with local connections, five line wide general purpose interconnect routed via switch matrices, and long lines – two per horizontal channel and three per vertical channel. The logic and I/O blocks are enhanced versions of the 2000 series cells.

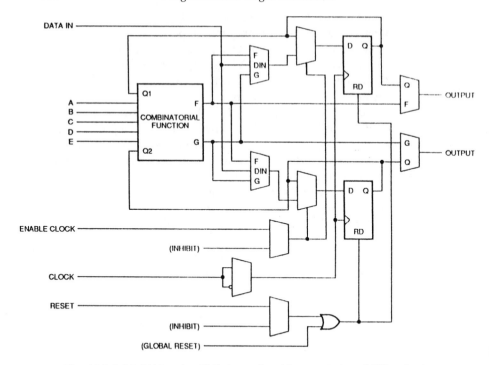

Fig. 6.22 LCA 3000 series CLB *(reproduced by permission of Xilinx Inc.)*.

6.2.3.2 3000 series logic block

Figure 6.22 shows the 3000 series CLB, which is an extended version of the 2000 series CLB. The additional resources are a second flip-flop and more inputs with a more powerful combinational logic function. This has five direct inputs from the interconnect plus feedback from both flip-flops. It can implement a five-variable function or two four-variable functions.

The flip-flops can select either of the combinational outputs or a direct data input. Apart from the clock, a clock enable and reset signals are also available from the interconnect. As with the 2000 series, the CLB outputs are either from the flip-flops or the combinational function.

6.2.3.3 3000 series I/O block

Registered outputs, as well as inputs, are provided in the 3000 series I/O cells. These have bypass options; other options are inverters for both output and output enable signals, slew rate selection and resistive pull-ups. Taken together with the CLB enhancements, these improvements make the 3000 series a more versatile and universal family.

6.2.3.4 3000 family devices

There are, currently, five members of the 3000 series. They range in size from 64 to 320 CLBs, the larger devices having rectangular rather than square arrays. The I/O ranges from 64 to 144 pins, and the largest device requires 64160 bits to define its logic function. A 65536-bit serial PROM has been introduced to cope with the larger RAM size in the more complex devices.

6.2.4 ERA family

6.2.4.1 ERA60100 architecture

This family uses a much simpler logic cell than the LCAs, although it is still a RAM-based device. The first member of the family, the ERA60100, contains 2500 logic cells in a 50×50 array. As with the LCAs, there are three types of interconnect, local, short range and long range. Each cell has eight inputs from a mixture of interconnect types; an internal dual four-to-one data selector picks two of these inputs for the cell logic function, which may be a simple gate function or latch. More complex functions are built up by connecting cells with the local connections.

6.2.4.2 ERA internal bus structure

The interconnect structure is shown in Figure 6.23. Local interconnect connects adjacent cells, as described above. The short range bus spans ten cells in either direction, while the long range bus covers the length or breadth of the whole chip. In addition, there is a 10-bit peripheral bus surrounding the whole chip. This is accessible to both I/O cells and logic cells, either as input or output to either. It can be used for distributing clock and reset signals round the array, as a data bus or for additional routeing.

6.2.4.3 ERA family

Larger devices are planned in an improved process up to a cell count of 20 000. These will form an ERA70000 series.

6.2.5 ACT1/TPC10 gate arrays

6.2.5.1 ACT1/TPC10 architecture

This family of gate arrays is the closest approach in programmable logic to masked arrays. Each cell in the array is a simple two-level logic function which may be programmed to emulate all the standard single-level logic functions, and many two-level functions. The cells are arranged in rows with horizontal routeing channels between each row.

Each routeing channel has 22 signal lines and 13 vertical lines across each cell. There is a potential 'anti-fuse' connection at every crossing point, plus

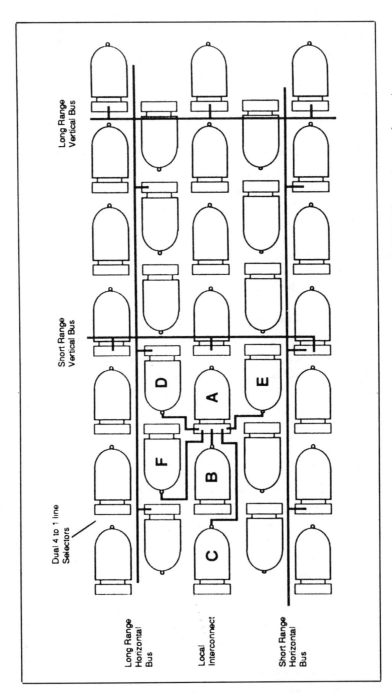

Fig. 6.23 ERA interconnect structure (*reproduced by permission of GEC Plessey Semiconductors*).

connections to the module itself, making an average 340 possible connections per cell. Given that there are 295 logic modules in the smallest array, there are over 100 000 potential connections in a full device. Only a small fraction of these would be used in any one design, but we can see how the small size of the anti-fuse technology is needed to make this range of devices feasible.

The small size of the anti-fuse also means that there is very little loading effect on the signal lines from unblown connections. With a series resistance of about 200 ohms, each blown connection adds a small predictable delay to the propagation time between modules. In practice, the number of connections is limited to four to avoid loading signals too much. This has only a minimal effect on routeability owing to the large number of routeing lines available.

6.2.5.2 *ACT1/TPC10 logic module*

The schematic of each logic module is shown in Figure 6.24. It contains three 2-input multiplexors, the output of two act as inputs to the third, and the select input of the third is derived from a 2-input OR gate. By connecting various of the eight inputs of the module to ground or V_{cc}, several different gating functions may be implemented. Figure 6.25 shows some examples based on a single multiplexor.

Examination of the gate level diagram for a single multiplexor, Figure 6.26, also shows how a latch may be implemented, simply by connecting the output to one input. Two modules in series, wired as latches, will thus form a D-type flip-flop, so this logic module can form the basis of any standard logic function.

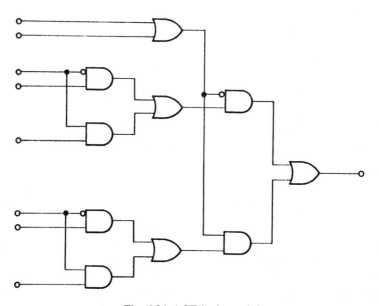

Fig. 6.24 ACT logic module.

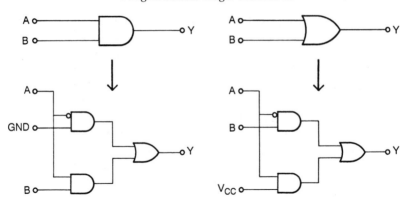

Fig. 6.25 Multiplexor gating functions.

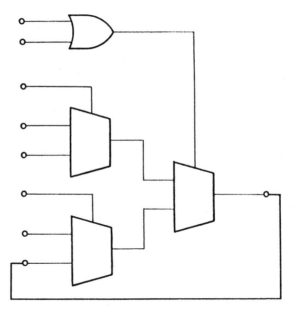

Fig. 6.26 Multiplexor wired as a latch.

6.2.5.3 ACT1/TPC10 I/O and range

There are just two devices in this range; their I/O availability depends partly on the package in which they are supplied. For example, the smaller 1010 device has 57 I/Os in total but it may be supplied in a 44-lead version with only 34 signal pins; this device has 295 logic modules. The larger 1020, with 547 logic modules, may have 69 external connections in an 84-lead package, but has the same I/O capability as the 1010 when used in its 44 or 68-lead versions.

Each I/O pin may take any of the usual I/O functions, input, output, tri-state or bidirectional.

Speed performance in this type of device architecture is best defined in terms of internal module performance, as propagation delay through the device depends on the number of logic levels needed to implement a particular function. Clock frequency is defined as 70 MHz and propagation delay through a single level macro is about 6–11 ns, depending on internal fan-out. I/O delays can add as much as 20 ns, again depending on the number of internal loads connected to the input buffer.

Power consumption depends on how many modules are used in a design, and how fast they are being run. A formula for calculating power is available; an example of an 85% populated 1020 device running at 16 MHz gives a result of about 90 mA for supply current.

6.2.6 ACT2/TPC12 series

6.2.6.1 Architecture and logic enhancements

The most obvious change made between the ACT1/TPC10 and ACT2/TPC12 families is the introduction of a second type of logic module. The original combinational module is termed a 'C-module'; the new module contains an additional latch/flip-flop element. As it can be used directly for sequential functions it has been called an 'S-module'.

To allow for this increased logic power per module, additional interconnections have been provided; there are now 36 horizontal tracks in each channel and 15 vertical tracks per logic module. This increased connectivity means more antifuses – as many as 750 000 in the largest device in the family.

6.2.6.2 Range and performance

Timing and delay specifications are very similar to the original family and, as before, a formula is given for calculating power consumption.

Three devices are planned for the family. The 1225 contains 231 S-modules, 220 C-modules and 82 I/Os; the 1240 has 104 signal connections and 684 logic modules, mostly S-types, while 1232 modules, of which 468 are combinational, are planned for the 1280, along with 140 I/Os.

It is claimed that over 200 standard TTL logic packages can be replaced by the 1280, which clearly takes programmable logic into the realm of VLSI (very large scale integration).

6.2.7 PA family

6.2.7.1 Architecture

Like the other FPGAs, the PA family has a set of logic modules set in an interconnection matrix. Unlike other FPGAs, though, the crossing points are

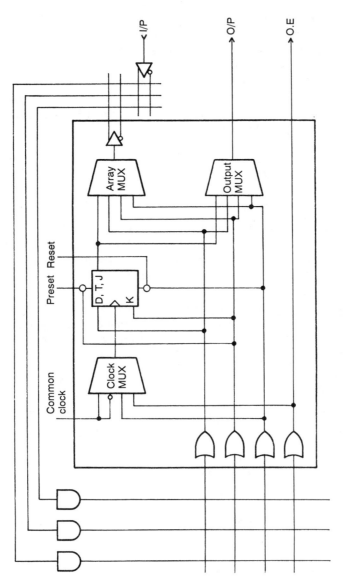

Fig. 6.27 PA7000 series logic control cell.

not simple interconnections but gating functions. In physical terms, the *x*-axis is a series of product term buses, the *y*-axis contains sum terms and the inputs to the product term array from the I/O cells and logic cells. In electrical terms, the area surrounding a logic cell can be simplified to look like Figure 6.27.

The array behaves as if it were a large PLA. Each device input and each logic cell has a true/complement input to the AND array, and each logic cell has four inputs from the OR array. The logic cells are properly known as *logic control cells*, and are the region of the diagram enclosed in a box.

The core of the cell is a universal flip-flop. The clock source, from one multiplexer, is the global clock, or its complement, or inputs 'C' or 'D'. Input 'D' may also be used as the output enable for the I/O cell associated with this logic cell. Input 'A' drives the flip-flop D/J/T input and 'B' is used for the K input in J–K mode. Set and reset may be global, or inputs 'B' and 'C' used for these functions. Finally, identical multiplexers select feedback and output signals from 'A', 'B', 'C' or the flip-flop output.

One I/O cell is associated with each logic cell and gives the options of true and complement controlled I/O, direct output, and direct or registered input. A global cell provides the control and configuration for the logic cells, arranged in groups.

6.2.7.2 Device descriptions

At present, two devices make up the PA family. The 24-pin PA7024 has 20 logic control cells, and I/O cells, leaving two pins which can be either clocks or inputs. With 22 potential inputs and 20 feedback signals, each AND term has 84 inputs; there are 80 AND terms in all driving 80 OR terms, four for each logic cell.

The larger PA7040, with 40 pins, has 24 logic and I/O cells, 12 dedicated inputs and two input/clock pins. There are 120 124-input product terms driving 96 sum terms. Both devices have their logic cells divided into two groups. It could be argued that these devices are really FPLSs, but their physical structure makes them more akin to FPGAs.

6.2.8 Programmable gate array summary

6.2.8.1 PGA performance

The logic capabilities and performance of the devices discussed in this section are summarised in Table 11. The supply current figures are, in all cases, for the stand-by mode; that is, d.c. conditions through the whole array. Actual supply currents depend on the toggle rate of each cell and the number of cells utilised in a particular design.

6.2.8.2 PGA/PLD comparison

Many of the PLDs described in Section 6.1 are also called FPGAs by their manufacturers. We have reserved the term FPGA for those devices which

Part number	I/O cells	Logic cells	Performance $f._{max}$	I_{cc}	Typical logic cell component structure
LCA2064	58	64	70 MHz	10 mA(SB)	4ip/2op std. logic + f/f
LCA2018	74	100	70 MHz	15 mA(SB)	4ip/2op std. logic + f/f
LCA3020	64	64	70 MHz	tba	5ip/2op std. logic + 2f/fs
LCA3030	80	100	70 MHz	tba	5ip/2op std. logic + 2f/fs
LCA3042	96	144	70 MHz	tba	5ip/2op std. logic + 2f/fs
LCA3064	120	224	70 MHz	tba	5ip/2op std. logic + 2f/fs
LCA3090	144	320	70 MHz	tba	5ip/2op std. logic + 2f/fs
ERA60100	84	2500	100 MHz	0.3 mA(SB)	2ip (from 8) latch/NAND gate
ERA70400	tba	10000	100 MHz	0.3 mA(SB)	2ip (from 8) latch/NAND gate
PA7024	20(+2ip)	20	41 MHz	140 mA(SB)	Reg. Cell/P-term interconnect
PA7040	24(+14ip)	24	41 MHz	155 mA(SB)	Reg. Cell/P-term interconnect
ACT/TPC1010A	57	295	70 MHz	10 mA(SB)	2 level triple 2ip mux
ACT/TPC1020A	69	547	70 MHz	10 mA(SB)	2 level triple 2ip mux
ACT/TPC1225	82	451	80 MHz	0.35 mA(SB)	triple 2ip mux + latch/f-f
ACT/TPC1240	104	684	80 MHz	0.35 mA(SB)	triple 2ip mux + latch/f-f
ACT/TPC1280	140	1232	80 MHz	0.35 mA(SB)	triple 2ip mux + latch/f-f

Table 11 Programmable gate arrays.

have the same structure as a mask programmable gate array. The key difference lies in the method of implementing the interconnect. In PLDs the connections between cells all pass through a central interconnection bus of some kind. The cells themselves are of the wide gating PAL type, the connections to each gate being defined by cells which are effectively within the gate itself.

FPGAs have a distributed interconnect which is normally available only to those cells which are physically adjacent to it. The cells themselves usually have only a small number of inputs, and both inputs and outputs are joined to the interconnection by cells or fuses which are external to the logic cell.

The FPGA arrangement has an advantage which may also prove to be a drawback. Multilevel logic, of the kind which is needed to make arithmetic and highly randomised logic, is easily implemented in an FPGA, and design is more akin to logic design with discrete logic circuits. Translation of existing discrete logic designs may therefore be more straightforward and more likely to succeed with FPGAs.

The drawback comes with speed performance. PLDs, in general, are very often used in interfacing circuits and, with the introduction of 16 and 32-bit processor architectures, wide gating is a distinct advantage. Wide gates have to be implemented in multiple levels in FPGAs, so there is likely to be a considerable speed penalty in comparison with conventional PLD structures. Luckily, as we shall see in the following chapters, the translation of design information into device structure is normally transparent to the user.

Chapter 7
Using Programmable Logic

7.1 WHEN TO USE PLDs

7.1.1 Economic considerations

7.1.1.1 Cost of ownership

Our investigation of the design process should start with a consideration of the criteria for using PLDs, rather than some other kind of device, for creating the logic circuit which we require. An early decision must involve the economic feasibility of this approach. To take the argument to an absurd limit, it would be possible to use a PLD in place of a simple gate, but to do so would involve four economic penalties. The PLD is more expensive, it would occupy more printed circuit board area, it would consume more power and it would require programming. How then do we set about finding the right economic situation for using a PLD?

One way to quantify the factors involved in this decision is by using a principle known as *cost of ownership*. When we buy any commodity the cost of owning it does not stop when we leave the saleroom. This applies to anything from a bag of flour to owning a house. Any householder knows that he has to pay taxes, repair bills, heating, decoration, furnishing, etc. in order to make a house habitable. Even converting a bag of flour to cakes involves the added costs of ingredients, cooking, washing up and, not least, the cook's time. If we buy integrated circuits we do this, presumably, in order to build some more elaborate system which we then hope to sell. Each component in that system adds to the cost of that system by virtue of being part of that system. The various factors in that added value can be analysed to find the most economic way of building that system.

Before an integrated circuit can be used it has to be purchased, checked on arrival and stored ready for use. At some time before going to the production floor it may be tested; if it is a PLD it will need programming, either by the manufacturer or the user. It will probably be paid for before any cash is recovered by selling the system it is used in, so interest will be lost on its cash value. Some wastage will occur in handling the devices by accidental damage, using the wrong device or many other human failings. All this will happen before it gets near a printed circuit board.

The cost of the printed circuit board itself is directly related to the size of the integrated circuit and the number of pins it has. The board will need testing and some boards will fail. Using PLDs can have one of two consequences. The number of boards may be reduced, in which case handling costs will be reduced proportionately, or each board will be smaller, which makes the chance of a fault occurring lower. In either case there is a cost which can be associated with each component. There is also a labour cost associated with manufacturing and a 'hardware' cost. That is the power supply, connectors and wiring loom as well as the box which the system is put in. Each component occupies space so more components means a bigger, and more expensive, box.

The final cost is concerned with problems after the system is delivered. Each component has a finite chance of failing, so each contributes to the expense of providing service under guarantee.

There are further costs which are less easy to quantify. Extensive use of PLDs shortens the development cycle, in itself a saving in engineers' salaries. The printed circuit board can be defined earlier, is easier to lay-out and less likely to need changing. The product can be brought to the market earlier and, if it proves more reliable, will enhance the manufacturer's reputation for quality thereby increasing sales. Thus using PLDs may prove beneficial even where accountancy cannot prove an immediate cost saving.

7.1.1.2 *Rentability calculation*

One method of finding the cheapest solution is to perform a *rentability* calculation; in it one estimates the actual cost of each factor and adds them together for comparison. A number of studies have been carried out in order to find the economic replacement value of PLDs (and other custom techniques). These all suffer from three assumptions which will be different in every case; what is being replaced, what is replacing it and what the user's costs are. In the following analysis we can specify the first two exactly, but can only guess the third. However each reader can put in the figures for their own situation and produce their own result.

We will calculate three sets of costs, an LSTTL SSI circuit, a medium PLD (e.g. GAL16V8) and a complex PLD (e.g. PAL22V10). It is intended that the figures are in pounds sterling, but relative costs in the UK and USA are such that US dollars may be equally applicable:

Costing category	LSTTL	MCPLD	HCPLD
Purchasing and stores	0.02	0.02	0.02
Testing and programming	0.05	0.12	0.14
Inventory and usage	0.03	0.10	0.18
PC board area	0.20	0.25	0.30
PC board fabrication	0.10	0.12	0.14
PC board test and rework	0.20	0.30	0.40
Power supplies and cooling	0.02	0.10	0.15
System hardware	0.05	0.06	0.07
Assembly labour	0.20	0.25	0.30
Servicing and reliability	0.07	0.07	0.07
Component cost	0.12	0.60	3.00
Total cost of ownership	1.06	1.99	4.77

We can, therefore, calculate how many LSTTL devices we need to replace to make it worthwhile using PLDs. Taking into account the other factors mentioned above it seems that a saving can be made by using a medium complexity PLD in place of two SSI devices, or by using a high complexity PLD in place of four or five. It must be emphasised that the above figures are estimates which will vary from application to application. However, the results are not untypical of the results obtained from most similar studies.

7.1.2 Technical considerations

7.1.2.1 Suitable application areas

If the economic analysis points towards PLDs the designer must still satisfy himself that the application is suitable. This is a technical decision. PLDs are not restricted to any particular sphere of application; every designer of electronic circuits should count them as a potential solution to his design problems. The alternatives to PLDs are discrete logic or masked ASICs and these too are universal in their application areas. The only reasons for not using PLDs are economic, or because they cannot provide an adequate performance.

7.1.2.2 Performance limitations

The technical limitations of PLDs are likely to be concerned with speed, power consumption or simply fitting the circuit into a single package.

The fastest PLDs compare well with the fastest discrete logic circuits. Although a 5 ns PAL is (at the time of writing) ten times the price of a 15 ns PAL, the performance of the 15 ns part may be equal to 5 ns discrete logic. This is because PLDs may replace two or more levels of logic, so the effect on the system will be the same as the discrete logic solution. Part of the reason for this multi-level replaceability is that the structure is based on an AND–OR architecture. Another factor is that PLDs use very wide gates, up to 20 inputs; to build a 20-input gate from discrete logic takes two levels of logic, hence the saving.

Now that CMOS is becoming more widely used for PLDs, the problems of high power consumption are becoming less. A typical gate package from a TTL family consumes less than 10 mA, while a low complexity bipolar PLD takes about 100 mA. A replacement factor of more than ten is needed to give a comparable power. The same function can now, probably, be achieved in a CMOS PLD with a quarter or half the power consumption of its bipolar equivalent. The penalties will be a slightly higher price, although this gap is narrowing, and slower speeds. The power consumption of CMOS increases with frequency anyway, so there is less to be gained in using CMOS at the highest speeds. True 'zero-power' is available for the simpler device structures, as we saw in Chapter 4.

With new structures being announced regularly, part of the designers problem is finding the best device to fit his circuit. The major constraint is the number of pins available in the package for I/O. If this is not enough for the circuit under consideration, more than one device may be necessary. In this case, the number of inputs is the critical factor, for it is usually easier to partition a circuit in a way which separates the outputs. The exceptions are state machines and other circuits which require outputs to be fed back to the array input.

Figure 7.1 lists most of the PLDs currently available. They are sorted into columns with the same number of inputs, and grouped according to the categories described previously in approximate order of complexity. The table may be used to select the best device for any application. Because input count is probably the most important factor, initial selection should be based on this. The column with the required number of inputs will contain appropriate devices.

A guide to the trade-off between inputs and outputs, for each device, is given by the bracketed figures below each device type. The upper line gives the range over which the input numbers span, while the lower line gives the corresponding information for outputs. Thus, if no device in the selected column has sufficient outputs, it should be possible to find one to fit the application in one of the higher-order columns.

INPUTS	8	9-10	11-12	13-14	15-16	17-18	19-20	21-22	23-27	28-39	40-63	64-99	>100
LSI PLDs					EPM5016 (0:...:10)			MAPL144 (2:..:21); PA7024 (2:..:21)(20..:1); MAPL128 (9:..:21)(11..:21)	EPM5032 (4:..:22)(16..:22)	MACH210 (5:..:27)(32..:77); MACH110 (4:..:37)(32..:1); ATV2500 (14..:97)(32..:1)	LCA3020 (13:..:57)(48..:1); EP1800s (18:..:63)(48..:1); EPM5128 (16:..:57)(48..:1)	LCA3042 (16:..:97)(96..:??); ER60100 (16:..:??)(48..:1); ACT1225 (61:..:83)(61:..:5)	LCA3090 (14a:..2); PGA2040 (7a:..2); ACT1280 (16a:..7)
SEQUENCERS	PLS159 (12:..:0)(12:..:4)	PLS157 (4:..:0)(12:..:6)	PLS168 (12)(8); PLS179 (8)(12:..:8); PLS155 (4:..:12)(12:..:8); 7C361 (8:...:12)	PLS167 (8); PLS506 (12)(8); PSG507 (12)(8)	PLS105 (8); PLUS405 (16)(8); PLC415 (16)(8)	30S16 (13:..:8)	GAL6001 (10:..:1)	42VA12 (19:..:21); 78C800 (12:..:21)(10:..:1)		PA7040 (14:..:37)(32:..:1); EPM5064 (8:..:30)(32:..:1); PGA2010 (55:..:??)	ATV5000 (61:..:??); LCA2064 (11:..:87); ACT1010 (44:..:95); EP900s (12:..:90); MACH220 (44:..:99)	LCA3030 (14:..:73); PGA2020 (9:..:74); LCA2018 (9:..:7); EPM5192 (9:..:71)	LCA3064 (11a:1?); ACT1240 (103:..?)
REGISTERED / **PLDs**	PAL16R8 (8)(8)	PAL16R6 (8:..:10)(8:..:12)	PAL16R4 (8:..:12)(8:..:12); 20X10 (10); PAL20R8 (8); PAL20X8 (12:..:10)(8:..:10)	PAL20R4 (12:..:14)(8:..:14)	PAL20R4 (12:..:16)(8:..:16); PAL20X4 (10:..:16)(8:..:16); GAL16V8 (10:..:16)(8:..:2); 16RA8 (8:..:16)(8:..:16); PAL23S8 (9:..:16)(8:..:1)	18(C)V8 (10:..:17)(8:..:1); EP320 (10:..:17)(8:..:1); 85C220 (10:..:17)(8:..:1); 18G8 (10:..:17)(8:..:1)	29MA16 (1:..:20)(8:..:20); 29M16 (10:..:20)(8:..:20); GAL20V8 (10:..:20)(8:..:20); 7B326 (12:..:20)(8:..:20); EP600s (16:..:20)(8:..:20); 85C060 (4:..:19)(8:..:20); 20RA10 (10:..:19)(10:..:1)	7C330 (19:..:20); 5AC312 (12:..:21)(12:..:1); 32VX10 (12:..:21)(10:..:1); 22V10 (12:..:21)(10:..:1); ATV750 (10:..:21)(12:..:21); 20CG10 (12:..:21)(10:..:1); 85C224 (12:..:21)(14:..:1)	7C332 (19:..:20); 26V12 (12:..:27); 7B333 (12:..:24)(10:..:1); 24V10 (12:..:24)(16:..:2); 7C331 (12:..:24)	5AC324 (24:..:39)(24:..:1); PLHS502 (24:..:??); PLHS501 (24:..:??)	MACH120 (32:..:63); PML2552 (32:..:??); PLHS601 (24:..:??)	EPM5130 (89:..:71); MACH230 (64:..:??); MACH130 (64:..:??); ACT1020 (44:..:99)	
COMBINATIONAL / **PLDs**		10L/H8 (10)(8)	12L10 (12); 12L/H6 (12)(6); 7B336/8 (12); 7B337/9 (12)(6)	PAL14L8 (14)(8); 14L/H4 (14)(4)	PLS100 (16)(8); PAL16L8 (10:..:16)(8:..:2); PAL16L6 (16)(6); 16L/H2 (16)(2); PAL16C1 (16)(2); PAD16N8 (10:..:16)(8:..:2)	PLS153 (8:..:17)(10:..:1); PAL18P8 (10:..:17)(8:..:2); PAL18L4 (18)(4); PAD18N8 (8:..:17)(8:..:2)	PAL20L8 (12:..:20)(8:..:2); 20L10 (12:..:20)(10:..:2)	PLS173 (12:..:21)(10:..:1); PAL20L2 (20)(2); 22P10 (12:..:21)(10:..:2); PAL20C1 (20)(1)	24L10 (16:..:24)(10:..:2)		48N22 (36:..:48)(22:..:10)		

Fig. 7.1 Summary of available PLDs.

7.2   A METHODICAL APPROACH TO PLD DESIGN

7.2.1   Outline of the steps

7.2.1.1   How to start!

There is an apocryphal story of the local resident who, on being asked directions, stated, 'If I were going there I wouldn't start from here.' So, how does one go about setting out on the path to a logic circuit?

Most systems take an input from the outside world, manipulate it in some way and then present the result to the outside world again. If the manipulation is carried out by a logic circuit then the information must be converted to a form which the logic can relate to, and then converted back to a form in which the outside can understand it. Examples are: computers, instrumentation, communications circuits and even PLD programmers. The conversions are the job of interfaces, the manipulations are done by the logic circuits. In many cases the main job of the logic can be undertaken by microprocessors or dedicated LSI circuits, which frequently need combinational logic to glue them together. Sometimes the whole logic function may be implemented by a collection of standard logic circuits.

In either case there is likely to be an area where PLDs may be used to advantage. Combinational PLDs more often fall into the glue category, while registered PLDs are frequently used to make broader-based functions. These boundaries are by no means rigid and both lend themselves to the same basic design techniques. In most cases, though, there will be some dedicated LSI involved so the design will start by defining these components. In the event of no dedicated LSI chip being available, it may be possible to design an LSI to perform the required function.

The next step is to decide how to connect the LSIs together, and generally some help is forthcoming from their data sheets to indicate which signals they need to work together. At this stage it is usually necessary to start inserting inverters, gates and more complex functions into signal paths. This is because control signals from one device are not exactly what is required by another, particularly if they come from different families of LSI. The end result is a circuit diagram containing a few LSI devices connected by several discrete logic devices.

7.2.1.2   Partitioning the circuit

If we have a circuit diagram the design has only just started. We have to decide which physical devices are best suited to actually implementing the logic function we have drawn. There will be constraints on this decision:

- How much time can we allow for signals to pass between devices?
- How big can we make the printed circuit board?
- How much power can we afford to dissipate?

Let us assume that the answers do not rule out PLDs, they may even make the use of PLDs mandatory. The first stage in designing the PLDs which we are proposing to use is a process called *circuit partitioning*. This is done by drawing a box round the parts of the circuit containing logic functions which are not incorporated into the LSI; we can now count the number of signals entering the box, the number of signals leaving the box and the number of gates or equivalent functions contained inside the box. The lower the sum of the first two numbers and the higher the third number, the more likely it is that a PLD would be the best solution. The chances are that with less than 25 inputs and outputs, and more than eight gates it will be worth considering a medium complexity PLD.

The justification for this is that 24 is the highest number of I/Os in a low or medium complexity PLD and more than eight gates probably means more than two discrete logic circuits, which we have seen is the minimum number worth replacing. If the number of I/Os is substantially higher than 24, either an LSI or more than one medium complexity device will be needed. In the latter case the partitioning needs to be done to make several boxes, all of which conform to the criteria above. If the number of gates is too low to make two or more PLDs economic then one solution might be to use a PLD to incorporate most of the logic, together with discrete logic to take care of the leftovers.

7.2.1.3   The design steps

We have reached the stage where we have one or more blocks of logic defined by logic symbols connected by signal paths. The next step is to convert this into a format suitable for entering this data into a PLD programmer. The most common way to define PLDs is by Boolean logic equations, so we should convert the logic diagram to equations. Starting from the inputs we can give each signal path in turn its logic equivalent until we arrive at the outputs. In this way we have defined each output as a logic combination of the input signals. Because PLDs are based on an AND–OR structure the equations have to be converted to the same format, which is often called *sum of products*. This is because the AND operator is manipulated in a similar way to multiplication, and the OR operator similarly to addition.

Conversion of the equation may be done manually although it is more usual to use a computer program; some software will also handle the logic diagram directly. At this point we can see how many AND gates are required for each of the outputs and an initial selection of PLD can be made. The factors which will influence this choice are:

* number of inputs
* number of outputs
* number of output signals fed back as inputs
* number of AND gates required for each output
* maximum delay time allowable
* maximum power consumption allowable

The procedure suggested in the previous section should indicate the preferred choice. If this does not give a simple PLD as the outcome, there are some steps which can be taken to try to simplify the logic. Firstly it may be worth trying a standard logic reduction technique, such as Karnaugh mapping described earlier in this book. This may reduce the number of gates required sufficiently to enable the use of a low complexity device and hence reduce cost.

A second consideration is the use of bidirectional pins. These may appear to be necessary if intermediate functions are formed from the input signals. If these functions are not needed as output signals they may be eliminated by expanding the equations involving them in the necessary outputs. This confers two benefits on the design. It may enable less or smaller PLDs to be used. Also there will be an improvement in delay time as the signals will not have to make two passes through the PLD.

7.2.1.4 *Completing the design*

Having reached the stage of possessing a minimised set of logic equations, and deciding on the target device, all that remains is to complete the design. This means that the logic information has to be mapped onto the fuse chart of the PLD. The method of doing this depends on the tools available to the designer. Software tools will be described in the next chapter, but whichever method is used the information will have to be entered into either a computer or programmer.

PLAs may usually be entered directly as a truth table so, without software, the problem facing the designer is to convert logic equations into a truth table. We addressed this problem in Section 4.2.3.1, but we may summarise the conclusion here. If the input variable appears in the equation without inversion then enter an 'H' in the truth table; if it appears inverted enter an 'L', otherwise, if it does not appear at all enter '–'. On the output side, if the output includes the equation enter 'A', otherwise enter '.'. If a bidirectional pin is being dedicated as an output all the inputs should contain '–'s, if it is an input all inputs should be entered as '0' otherwise the enabling equation should be entered for true tri-state operation.

The last question to be settled is, will the design perform the function which is intended. This can usually be solved by defining test *vectors* which are used to simulate the device performance, but need a computer program to make them operate. If this step is successful, the test vectors can be used to test real devices functionally, as we shall see later in this chapter.

7.3 PROGRAMMING

7.3.1 PAL/PLA fuse mapping

7.3.1.1 *Programming map*

We saw, in Chapter 3, how PROMs use the same addressing circuitry to enable the fuses for both programming and reading. This is because the input lines are fully decoded in the AND array. PALs and PLAs have a programmable AND array so the input lines cannot be used for decoding the fuses directly. Additional circuitry is needed to address the fuses and this is enabled by applying supervoltages to specified pins. The input pins then become equivalent to PROM inputs and address the fuse array as if it were a PROM. By applying supervoltages to the supply and outputs the programming circuitry itself is enabled.

Because the fuses are programmed as a PROM they are given a pseudo-PROM structure for programming purposes. This address map only refers to the programming situation and need not concern the design situation as it should be taken care of by the design software and programming equipment.

7.3.1.2 JEDEC fuse map

The standard way of transferring fuse information to a programmer is by the JEDEC fuse map. JEDEC is the Joint Electronic Devices Engineering Council; it has the task of ensuring that specifications for the same device from different manufacturers are compatible. This makes certain that a design completed on one manufacturer's device can be transferred to another manufacturer without changing the design parameters. It also gives producers of design software a standard output format.

JEDEC fuse numbers are not the same as the fuse address, which is the binary code used by the programmer. They run in sequence from '1' to the total number of fuses within the device and, usually, follow the order which is specified by the internal fuse map. A full JEDEC file contains a complete description of the device being programmed, including a list of blown fuses. A typical JEDEC file might appear as below:

STX (i.e. control + B)	
P.L. Systems Inc	(Company Name)
A. C. Guy	(Designer)
30 06 86	(Date)
ABC9876	(Drawing Number)
PAL16L8	(Part Number)
*	
*D2029	(Device Code)
*F0	(Default Fuse State)
*G0	(Security Fuse State)
*L0000 1111111111111111111111111111111111	(Fuse Information)
*L0032 101011110111011110010111101001111	(Fuse Information)
*L0256 1111111111111111111111111111111111	(Fuse Information)
.	
.	
*L1920 11101110101001110110111011010101	(Fuse Information)
*C4AB7	(Fuse Checksum)
*V0001 100111000N1HLLLHHXXN	(Test Vector)
*V0002 110011011N0LLLHXXLHN	(Test Vector)

*V0100 011100101N0HHHHLLLLN   (Test Vector)

*

ETX (i.e. control + C)

B642   (Transmission Check-
                                                sum)

Commercial programmers will accept data in the JEDEC format and translate it into a fuse map corresponding to the device structure in programming mode.

7.3.2   Device programmability

7.3.2.1   Intrinsic yield

The one part of a PLD which cannot be checked by manufacturers using metal or diode fuses is the programming circuit; any attempt to test it will cause part of the device to be programmed and therefore rendered useless. On the other hand, UV or electrically erasable cells can be programmed and then erased so it is possible to check that these devices are programmable before releasing them for delivery. The intrinsic yield of these devices, which tend to be MOS or CMOS, is likely to be higher than that of bipolar PLDs.

As with bipolar PROMs, the manufacturers of bipolar devices usually include extra fuses, which play no part in normal device operation, to test programmability. This, again, has the spin-off of allowing testing of the programmed output level which could not be guaranteed otherwise. Other checks to ensure good programmability are carried out during the manufacturing process. These include resistivity and thickness of the deposited metal layers, width of the fuses after etching and visual inspection of the circuits for damage to diffusion and metal layers after etching. Advances in manufacturing techniques such as ion implantation, plasma etching and electron beam lithography also reduce defect densities and contribute to higher programming yields.

7.3.2.2   Batch effects

Batch effects may occur due to manufacturing defects or to statistical fluctuations in the received devices. Considering manufacturing effects first, there are several areas where a defect might cause a batch problem with programmability. This starts at the design stage where a single mistake could cause a device to be over-sensitive to one of the process parameters; modern CAE techniques make this event unlikely. A single mask fault might cause the circuit in the same position on every wafer to be defective, while an out of tolerance etching or layer deposition stage could cause a whole batch to be marginal. The effect of this on the overall programming yield depends on the number of defective circuits per batch and the number of bad batches in the production. It is the task of the manufacturer's quality control department to prevent these faults reaching the user. Bad batches will tend to get diluted by good product in the subsequent stages of assembly, testing and distribution and appear as an overall programming yield to the user.

The actual number of rejects seen by the user depends on the overall proportion of defective devices and statistical fluctuations. For example, if the overall rate is 1% then normally no rejects would be seen if just ten devices were programmed but just occasionally, about nine times in every 100, one reject will occur. Less often, once out of 200 times, two rejects will be found in ten which can give a misleading impression of the true rate on the odd occasion when it happens. The larger the batch the more likely is the observed failure rate to be close to the true rate, as the following table shows:

TRUE RATE – 2%							
			Observed rate				
Batch size	*0%*	*1%*	*2%*	*3%*	*4%*	*5%*	*6%*
100	0.13	0.27	0.27	0.18	0.09	0.04	0.01
300	0.02	0.26	0.46	0.21	0.04	0.01	
1000		0.06	0.73	0.20	0.01		
3000		0.02	0.94	0.04			

The figure in each column is the probability of observing a failure rate centred on the value at the head of the column. Thus, in a batch of 100 devices there is a probability of 0.18 of finding three rejects; in a batch of 1000 there is a probability of 0.06 of finding from 5 to 14 rejects. As might be expected there is a lower chance of finding an excessive reject rate in larger batches, although the possibility of statistical fluctuations must be considered if a high reject rate is found in a batch of PLDs.

7.3.2.3 User problems

If devices are not programmed properly then, however good the quality of the product, the yield will be poor. In this section we examine the problems which a user can bring on himself because he is using poorly maintained equipment for programming. The most important factor is to ensure that the equipment meets the specification laid down by the device manufacturer. Every aspect of the specification is normally included to an exact tolerance and if the tolerance is exceeded there can be a damaging effect on the yield. The parameters specified will include pin voltages, current capability of the power supplies, width and edge speeds of pulses, relative timings of pulses, average power consumption and allowable number of programming attempts.

If the tolerances are exceeded the programming yield may suffer, as we saw above; worse though the device may be programmed in an unreliable manner. It is worse to use unreliable devices than to waste PLDs needlessly through misprogramming. The latter costs the price of the device while the former may cause equipment to fail in the field, which involves repair costs, and may result in lost reputation. If yields are lower than expected the programming equipment should be calibrated to eliminate the possibility of drift in the equipment.

If the electrical parameters are all in specification it is still possible for the equipment to be causing extra failures. The other critical part of the programmer is the socket. If contact between the socket and device pins is poor there can be two ways in which performance is likely to be affected. As we have seen before, a relatively high current is needed to blow the fuses in bipolar PLDs. A high resistance between socket and device pins can cause the voltage to drop below the level needed to turn on the programming circuit in the PLD. The most likely result is to cause a 'no blow' when the device fails to programme but is not otherwise damaged.

If there is poor contact on one of the addressing pins it is possible that the device will 'see' a HIGH although the programmer has put a LOW onto the socket pin. This will cause the wrong fuse to be programmed and may result in the device failing to verify after the programming sequence. If a HIGH is misread every time a LOW is presented then half the fuses will be programmed at the wrong time; the chance that this will not result in failure is negligible.

Most programmers use *zero insertion force* sockets, which allow the device to be dropped in and then contact is made by clamping the pins in the socket. These sockets can age by the cam wearing so that the clamping is ineffective, or by the plating on the contacts wearing or becoming contaminated. Regular cleaning is an obvious remedy; in addition the sockets should be changed regularly before they start to wear out. After all, if a socket costing £10 (or $10) is changed when 10 000 devices have been programmed the cost is only 0.1p (0.1c) per device. This is paid for by a yield improvement of 0.05% on a £2 ($2) device.

7.4 TESTING

7.4.1 Why test?

7.4.1.1 *Potential PLD faults*

We noted in the previous section that PALs and PLAs need a separate fusing circuit in order to address the fuse array as a pseudo-PROM. A programmer will usually follow a standard sequence during programming, as we saw before. The final stage in the sequence is a verification that the correct fuses have been blown, but the fuse addressing circuit has to be used for this. Faults in this circuit will be discovered in the verification but the possibility remains that there are faults in the 'logic part' of the PLD. This argument does not apply to PROMs as these use the same circuit for fuse addressing during programming and during normal operation.

The effect of this is not hard to imagine; having programmed a batch of PLDs and inserted them into his circuit, the user will find that some of the circuits do not work as he expected. This is because the PLDs are not following the correct logic pattern even though all the fuses appear to be correctly blown. This could be due to a fault in the fuse addressing circuit, but it is more likely to be because there is a defect in the logic path which has not been detected by the fuse verification. The solution is to test the logic path independently.

7.4.1.2 Test classes

Testing is a global concept when applied to any product and to understand the problems we need to define test classes. As far as this discussion goes we are interested only in *zero-hour tests*. These, as their name implies, refer to the performance of devices when they are unused. *Life tests*, on the other hand, measure how reliable devices are by seeing how long it takes them to fail, failure usually being measured as an inability to pass the zero-hour tests.

Zero-hour tests themselves can be divided into two groups, *parametric* and *functional* tests. Parametric tests usually measure the ability of the device to interface to other parts of the system, covering such features as the current taken by the inputs, current drive available from the outputs, voltage levels defining HIGH and LOW logic states and power consumption. These are often called d.c. tests. Also included in parametric tests are a.c. tests which measure timings relevant to circuit operation, such as propagation delays, setup and hold times and clocking speeds.

Parametric test results are not usually affected to a large extent by programming. There may be a small reduction in power consumption if some AND terms are removed, and this may also cause some speed improvement. On the whole though, any parameters measured and guaranteed by the manufacturer should not need further testing. We noted before that the test row and column are used to set the device to its active output condition; this allows the manufacturer to carry out full parametric testing.

Functional testing is another matter. An unprogrammed PLD will be completely inactive, apart from the test fuses, so there is no way of knowing whether the logic path is working fully. Whatever pattern of inputs is applied, the outputs will not change so there is no way of applying a functional test until after programming. A functional test, at its simplest, consists of applying different patterns of HIGHs and LOWs to the inputs, and checking that the correct pattern is obtained from the outputs. The input patterns are called *test vectors* and the skill in designing functional tests is to create an optimum set of test vectors.

7.4.2 Test coverage and testability

7.4.2.1 Design for test

The first criterion in designing a test program is whether the circuit to be tested is testable or not. At first sight it might appear that any circuit must be testable, but this is not necessarily the case. Consider the simple D-latch; the simplest equation is:

$$Q = D * LE$$
$$+ Q * /LE$$

This is testable, provided that we start with LE HIGH, because a fault in any of the three gates can be detected. Every combination of HIGHs and LOWs can be applied to each gate and the resulting output verified without interference from any other gate. However, to achieve a glitch-free output, we modified the design to:

$$Q = D * LE$$
$$+ Q * /LE$$
$$+ D * Q$$

Now we find that some of the gates interfere with each other. In particular we cannot verify the action of the new AND gate which we have added, not surprisingly as it does not alter the function of the circuit. In order to make the circuit fully testable we would have to add a test control signal 'T' which could isolate the third gate. The testable glitch-free circuit becomes:

$$Q = D * LE * T$$
$$+ Q * /LE * T$$
$$+ D * Q$$

Taking 'T' LOW will disable the first two gates and enable the third gate to be tested in isolation, having first set Q to a known level. This is an example of a circuit being untestable because of parallel logic paths; this can be resolved by disabling one path for the test sequence.

A second cause of untestability is starting in an unknown state. For example, a simple counter will increment its count by one for each clock pulse it receives, so part of the test sequence will be to block the counter and check for an increase of one. If we do not know what the first count state is we do not know which state to check for after the clock pulse. The remedy is equally obvious; provide a means of setting the counter to a known state, if necessary by supplying an additional reset signal for test purposes only.

A similar problem is that of illegal states. A decade counter, for instance, will not normally use output combinations corresponding to the hexadecimal numbers A–F, but it may be possible for it to enter those states accidentally; a parallel load function might allow these numbers to be entered. If no transition path is available out of these states the counter will be stuck there indefinitely. It is therefore advisable to provide a state jump for all 'illegal' states to a known state. The complement term is a convenient method of doing this in those devices which possess that feature.

The fourth cause of untestability which we shall examine is the case where there is a dead-end in the state diagram. It is difficult to envisage this being designed in deliberately, but it could occur accidently by a poor design allowing a particular input condition to trigger two jumps at once. If there is an inadvertent dead-end then, again, once entered the device will become 'stuck'. In particular, care should be taken when using the 'hold' type of feedback with D-register PLDs, for a hold is a dead-end unless some way of releasing it is included. The solution, as in the previous case, is to provide a default jump out of illegal states.

7.4.2.2 *Fault grading*

We have discussed the need for testing to find faults and we will now proceed to describe in more detail how to define them, to ensure that the test routines will check every possible fault. If we look at the simplest logic device, the inverter, we can see how faults may be described. When functioning correctly, the output from an inverter will be the opposite sense from the input; a faulty inverter will have an output which is either always HIGH or always LOW. These faults are referred to as *stuck-at-one* or *stuck-at-zero*, SA1 or SA0 for short. There is a third possibility, that the output is the same as the input. This condition is most unlikely as most faults are due to open circuits or short circuits to one of the supply rails; fault analysis is much easier to handle if only SA1 and SA0 faults are considered.

Fault grading is the process of finding how many of the possible SA1 and SA0 faults are detected by the test vectors being graded. Applying a HIGH to the input of an inverter will detect an SA1 output, for if the output is SA1 the test will fail. To detect an SA0 output as well it is necessary to apply a LOW to the input, for then we expect the output to go HIGH. We can illustrate this by looking at the old example of the simple combination lock. This had the equation:

$$OPEN = A * /B * /C * D$$
$$+ A * /B * C * /D$$

Figure 7.2 shows how this would be programmed in a simple PLA. It shows that there are 9 possible nodes in the circuit, each of which must be tested for SA1 and SA0 faults. The test vestor A–H; B–L: C–L: D–H tests lines 1,2,4,5,7,9 for SA0, for if any of these is stuck LOW the output will be LOW

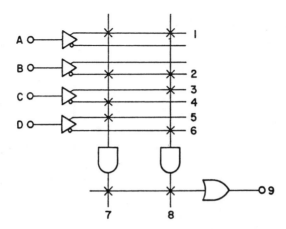

Fig. 7.2 PLA programmed with simple combination lock.

instead of HIGH. We can write out a full table of vectors to cover all possible faults in this circuit:

A	B	C	D	OPEN	
H	L	L	H	HIGH	SA0–1,2,4,5,7,9
H	L	H	L	HIGH	SA0–1,2,3,6,8,9
L	L	L	H	LOW	SA–1,7,8,9
H	H	H	L	LOW	SA1–2,7,8,9
H	L	H	H	LOW	SA1–4,6,7,8,9
H	L	L	L	LOW	SA1–3,5,7,8,9

Note that there are 16 possible test vector combinations from the four inputs, but that we have covered all possible faults with just six of these. This is an important fact because it means that a circuit with 'n' inputs may be fully tested with many fewer than the 2^n possible combinations of test vector. In PLD circuits 'n' can be as high as 20 which gives over a million possible combinations.

7.4.3 Designing test sequences

7.4.3.1 Simple combinational circuits

We can use the result from Section 7.4.2.2 to see the principles behind designing test sequences. To start with we can look at the simplest circuits, that is those using just combinational logic without feedback. Once again we will use the Karnaugh map as a tool; this time it will help us define the test vectors. Figure 7.3 shows the map for the logic function of the combination lock and a map on which the test vectors are plotted. The first two vectors correspond to the logic function and, in effect, check that the circuit responds to its active logic inputs. In other words, it ensures that the output goes HIGH when it should; these are sometimes called type-1 tests.

The other four vectors test the condition that the output stays LOW when the inputs are not in an active combination. That is that the output does not go

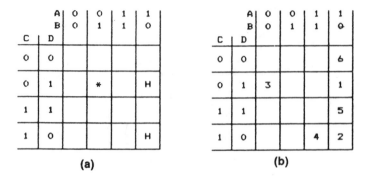

Fig. 7.3 Karnaugh maps: (a) combination lock; (b) test vectors.

HIGH when it should not, often referred to as type-2 tests. The Karnaugh maps show that the type-2 vectors are all adjacent to the type-1 vectors. This ensures that a fault on one input will not be masked by other inputs. If the starred cell was used as a vector and line 1 (Figure 6.2) was SA1 the test would still pass because input B is in an inactive state. The vector would not detect SA1 faults for either line 1 or line 2 unless, of course, both lines were faulty.

The Karnaugh map is useful in the mechanical process of defining output levels corresponding to input vectors in more complex systems. Most circuits will have several outputs and each output must be fully tested for SA1 and SA0 faults; often one input vector will be relevant to more than one output and more coverage will be obtained if every output is checked for each input combination. As an example of a more complex circuit we will derive the test vectors for a 4-input priority encoder. The PLA circuit and Karnaugh maps are shown in Figure 7.4. Input vectors are conventionally indicated by '1'

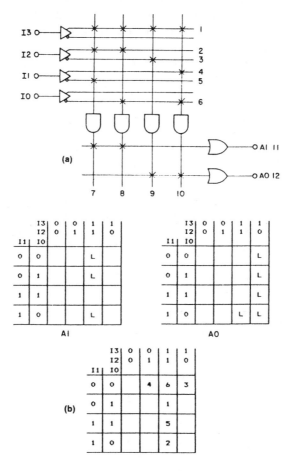

Fig. 7.4 4-input priority encoder: (a) PLA circuit; (b) Karnaugh maps.

(HIGH) and '0' (LOW), while the resulting outputs are given 'H' and 'L'. The table of test vectors is:

I3	I2	I1	I0	A1	A0	
1	1	0	1	L	H	SA0–1,2,5,7,11 SA1–3,9,10,12
1	1	1	0	L	L	SA0–1,2,4,6,8,10,11,12
1	0	0	0	H	L	SA0–1,3,9,12 SA1–2,7,8,11
0	1	0	0	H	H	SA1–1,7,8,9,10,11,12
1	1	1	1	H	H	SA1–3,5,6,7,8,9,10,11,12
1	1	0	0	L	H	SA1–3,4,9,10,12

Various points are illustrated by this example. Vectors 1 and 3 can cover both a type-1 and a type-2 test for the respective outputs; every input is defined for each vector even though some inputs are 'don't care' for some outputs; vector 6 is not suitable for a SA0 test on A1 because this is an overlapping cell which falls into the parallel logic path category of untestability. Once again we need only 6 out of the 16 possible combinations to cover all the faults.

7.4.3.2   Combinational feedback circuits

The same principles apply to circuits with feedback as to simple combinational circuits, but there is the added complication that the level of the feedback input may need to be defined by other inputs. An alternative approach is to break the feedback, if possible, by using the tri-state control or a test enable input. This may not be possible if 'spare' inputs are unavailable. The problem is not so acute when the feedback is used to create logic circuits with several levels, as with a parity generator, which may have a number of exclusive-OR gates cascaded. These may be tested as in Section 7.4.3.1, but circuits such as the D-latch must be treated as sequential circuits. We can see this if we derive the test vectors for the 'testable glitch-free' D-latch from Section 7.4.2.2. The Karnaugh maps and PLA circuit are shown in Figure 7.5 and yield the following vectors:

D	LE	T	Q	
1	1	1	H	SA0–9
1	0	1	H	SA0–9
0	0	1	H	SA0–3,4,5,7,9
0	1	1	L	SA1–1,6,9
0	0	1	L	SA1–5,7,9
1	0	0	L	SA1–1,8,9
1	1	1	H	SA0–1,2,4,6,9
1	1	0	H	SA0–1,5,8,9
0	1	0	L	SA1–4,6,9

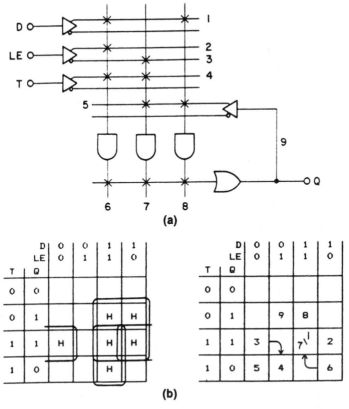

Fig. 7.5 D-latch: (a) PLA circuit; (b) Karnaugh maps (function and test vectors).

Because of the feedback we are unable to isolate LE so that lines 2 and 3 cannot be tested for SA1 faults. It may be seen from the Karnaugh map that the test progress in sequence round the map as one input at a time changes. The arrows show where the output sense changes with a consequent change in map position, again because of the feedback. By changing only one input at a time we avoid timing problems. If we took LE LOW and changed the sense of D in the same test then the output level would depend on which change occurred first. This race condition should be avoided in designing test sequences as much as in designing the circuits themselves.

7.4.3.3 Sequential circuits

We will assume that by sequential circuits we mean state machines. Test sequences for state machines can become very complex; usually it is best to consider transitions from each state in turn, in which case it may be possible to treat each state in a similar fashion to a combinational circuit. Type-1 tests are then those which are designed to cause a state jump, while type-2 tests are those which should leave the state register unchanged. The main complication occurs with devices containing buried registers, although in some cases it is possible to read these via output pins by applying a supervoltage to a specified input.

Rather than base the test sequence on an analysis of SA0 and SA1 faults of the internal components it is usually simpler to use the state diagram as a basis. In practice the result will be almost the same and will have the advantage that it will probably mirror the final application more exactly. In Figure 7.6 we

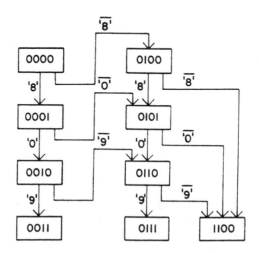

Fig. 7.6 Enhanced combination lock state diagram.

have reproduced the state diagram for the enhanced combination lock in order to show how to derive a set of test vectors. The first step is to enter a known state, by means of the asynchronous reset, and then run through the states with 'correct' entries. After that we must check that false entries take us into the error states; we still use the SA0 and SA1 principle by having only one bit at a time false when moving into errors. The full table is:

CK	RS	I3	I2	I1	I0	Q3	Q2	Q1	Q0	UN
0	1	1	1	1	1	L	L	L	L	L
C	0	1	0	0	0	L	L	L	H	L
C	0	0	0	0	0	L	L	H	L	L
C	0	1	0	0	1	L	L	H	H	H
0	1	0	0	0	0	L	L	L	L	L
C	0	0	0	0	0	L	H	L	L	L
C	0	1	0	0	0	L	H	L	H	L
C	0	0	0	0	0	L	H	H	L	L
C	0	1	0	0	1	L	H	H	H	H
0	1	1	1	1	0	L	L	L	L	L

CK	RS	I3	I2	I1	I0	Q3	Q2	Q1	Q0	UN
C	0	1	1	0	0	L	H	L	L	L
C	0	1	0	1	0	H	H	L	L	L
0	1	1	1	0	1	L	L	L	L	L
C	0	1	0	0	1	L	H	L	L	L
C	0	1	0	0	0	L	H	L	H	L
C	0	1	0	0	0	H	H	L	H	L
0	1	1	0	1	1	L	L	L	L	L
C	0	1	0	0	0	L	L	L	H	L
C	0	0	1	0	0	L	H	L	H	L
C	0	0	0	1	0	H	H	L	H	L
0	1	0	1	1	1	L	L	L	L	L
C	0	1	0	0	0	L	L	L	H	L
C	0	0	0	0	1	L	H	L	H	L
C	0	0	0	0	0	L	H	H	L	L
C	0	1	0	0	0	H	H	H	L	L
0	1	1	1	1	0	L	L	L	L	L
C	0	1	0	0	0	L	L	L	H	L
C	0	0	0	0	0	L	L	H	L	L
C	0	1	0	1	1	L	H	H	L	L
C	0	1	1	0	1	H	H	H	L	L
0	1	0	1	1	1	L	L	L	L	L
C	0	1	0	0	0	L	L	L	H	L
C	0	0	0	0	0	L	L	H	L	L
C	0	0	0	0	1	L	H	H	L	L

The full test sequence thus takes 34 vectors. One reason for this is that every state has to be tested individually as a simple combinational circuit, so the number of tests per circuit has to be multiplied by the number of states to find the total number of tests. The reason is that once a state has changed it must be re-entered before it can be tested again. Part of the above sequence is taken up by moving round the state diagram back to test the appropriate state with another input. Some PLDs have the ability to load the output register by means of a preload function or input supervoltage; the need for such a facility is amply demonstrated by the test sequence we have just created.

7.4.4 Summary of test design procedure

7.4.4.1 Pitfalls

Testing is such an important stage in designing PLDs that we make no apologies for summarising the above discussion. Pitfalls to be avoided as part of design for test and test design are:

(1) parallel logic paths;

(2) undefined entry states;
(3) dead-ends (states with no exit);
(4) race conditions where the output depends on the order in which signals are applied;
(5) floating inputs (unless they are not used by ANY AND term).

7.4.4.2 Procedures

The following comments can apply to either combinational or sequential circuits bearing in mind the principle that sequential circuits may be tested by treating each state as a separate combinational circuit. In order to obtain the fullest possible test coverage, these procedures must be followed:

(1) Test type-1 faults by enabling one AND-term at a time.
(2) Test type-2 faults by negating one input at a time for each AND term.
(3) Define the expected level of every output for each input vector.
(4) Use register preload wherever possible.

Chapter 8
Support For PLD Users

8.1 PLD USER NEEDS

8.1.1 Software requirements

So far we have looked at PLD structures and the ways in which their logic content can be specified. There is still a gap to be bridged though; that is to establish the correspondence between connections to the logic gates in the circuit diagram, and the actual fuses or cells in the device itself. This is usually a job for a computer program, although in some cases the device programmer can accept manual input in a format which can be translated directly into a fuse map. An example is truth table format for some PLA devices.

We will look at detail in the next section, but it is worth noting first the functions which may be required from a software package. The primary interfaces are from the designer and out to the programming equipment. The program must be able to determine which fuse or cell locations in the fuse map are to be programmed and produce an output, probably in JEDEC format to provide the output interface.

A second important function is logic optimisation. Logic designers are primarily concerned with ensuring that the final device will perform all the functions which are needed from it. In doing this they may well duplicate terms or overlook simplifications which could be made to the logic definition. This could result in an unnecessarily complex device being specified. A logic optimisation routine should reduce the designers input to its most compact form, enabling it to fit into the most economic device available.

We have already discussed vector testing of PLDs; the software equivalent of this is simulation. In a simulation package the designer can define a sequence of logic levels to be applied to the device inputs; the software will then predict the output logic levels resulting, or compare them with the designer's own predictions or requirements. It will then, usually, go on to generate the vector test sequence for testing programmed devices.

A final feature to be considered is documentation. Having completed the design of a system, or piece of equipment, a designer will move on to a new system or, perhaps, a new company. At some time in the future he, or his successor, may well be faced with a problem in the design. Unless the original design is well documented, much time could be wasted in trying to figure out exactly what the various parts of the design were trying to achieve. A well documented design must, therefore, be possible with any PLD design software.

8.1.2 Hardware requirements

A device programmer must accept the PLD fuse map and programme the fuses or cells in the target device according to that information. The basic requirements of a programmer are thus a communications interface, sufficient memory to store the fuse map and the means of applying the programming signals with the specified pulse widths, delay times and voltage and current levels.

In addition, desirable features are the ability to view and edit fuse maps, to blow the security fuse and perform vector testing. We can expand on these requirements, and the way in which they are provided in Section 8.3.

8.2 DESIGN SOFTWARE

8.2.1 Resident hardware

The first thing to establish with a computer program is the hardware and operating system needed to run it. The universal hardware, at present, is the IBM-compatible personal computer running under MSDOS or PCDOS. These are not the only machines available; newer versions using 80386 or 80486 processors may run on a new operating system called OS/2, and there is a substantial market for competitive machines such as the Macintosh. Other variables in the 'standard' PC range such as RAM size, disk drive configuration and monitor resolution may affect the ability to run certain programs.

Other machines which are commonly used for circuit design in general are *work stations*. These are high speed computers running dedicated design packages with large memories and high resolution screens, and which have only a very basic operating system as they are not intended for general purpose computing.

Compatibility with any existing design or general purpose hardware should therefore be established, otherwise new hardware as well as new software may have to be purchased. A final parameter to establish is operating speed. Some features such as logic optimisation may be unacceptably slow when run on a basic machine so, again, new hardware could be required in order to make full use of some design packages.

8.2.2 Data input methods

8.2.2.1 Logic equation entry

We have already seen how logic circuits may be defined by Boolean equations. In practice, this is the most usual method of entering data into PLD logic compilers and every design program in common use has this facility. There are format and syntax differences between many of the programs, which means that files are not usually interchangeable between them. While this is not the place to list detailed differences between the programs, we can point out the areas where care should be taken to ensure that the correct format is used.

Apart from listing information about the design, such as the designer's name and company and the description and revision of the design, the first data to be defined is, usually, the assignment of signal names to pin numbers, and the target device. Some programs do not require this information until the full logic content has been defined, thereby giving the designer freedom to leave the choice of target until the full implications of the logic requirements are seen.

Logic equations take the format:

OUTPUT SIGNAL = FUNCTION OF INPUTS AND FEEDBACK

The first differences arise in the way in which active-LOW outputs and inversions are defined. Common alternatives for an inversion are "/" and "!". Some programs require the signal name to be inverted, so that a signal defined as OUTPUT is entered as /OUTPUT (or !OUTPUT), while a signal defined as /OUTPUT is entered as OUTPUT. This can cause problems when the output signal is used as a feedback, when care has to be exercised as to whether it should be preceded by an inversion or not.

The alternative method of defining output inversion is to invert the whole logic function, thus:

OUTPUT SIGNAL = /(FUNCTION OF INPUTS AND FEEDBACK)

This tends to be safer in the way in which feedback levels get defined, but some programs, if confronted with an expression of this form, will expand it into an active-HIGH expression. This may not be a problem, depending on the target device envisaged and on which solution uses up less product terms.

The second common discrepancy between programs lies in the choice of symbol for the standard logic functions. In this book we have standardised on '*' for AND, '+' for OR and : +: for exclusive-OR; some widely used programs use '&' for AND, '#' for OR and '$' for exclusive-OR. The latter course frees the '*' and '+' symbols for use as arithmetic operators for programs which support arithmetic functions.

Care must also be taken when defining registered outputs. Common practice was to use '=' for combinational equality and ':=' for registered, the latter implying equality after the next clock pulse. As we have seen, though, many devices now allow a choice of register type, so a more precise format has been generally adopted. Each input to a macrocell is defined by an extension to the signal name; some examples are:

OUTPUT.D = ... (D-type input)
OUTPUT.J = ... (J input of J–K flip-flop)
OUTPUT.OE = ... (output enable condition)
OUTPUT.RST = ... (flip-flop reset input)

Again, though, each program may have a different syntax from others on some of the options.

Finally, attention should be paid to the way in which the program decides that the equation for a given output signal is complete. Some demand a delimiting character, such as a ';', at the end of each equation; others detect the '=' at the start of the next equation, and assume that the previous equation has finished. Syntax errors are usually picked up during the logic compilation, meaning that corrections to the file are required; clearly, though, much time and frustration can be avoided by making sure of the syntax in the first place.

One of the drawbacks of logic equation entry is the amount of 'secretarial' work needed to complete a design. Some of the ways in which this can be reduced are discussed later, but one particular method of shorthand entry can be described now.

Where a group of signals forms a set with a well defined relationship, some programs allow the set and the relationship to be defined in an abbreviated format. For example, the notation:

[inp7 . . . 0]

represents the eight signals, inp7, inp6, inp5, inp4, inp3, inp2, inp1 and inp0. An equation may be written in terms of this set of inputs by, for example, assigning them a hexadecimal value as in:

OUTPUT = [inp7 . . .0] = = A6

This is another way of writing:

OUTPUT = inp7 * /inp6 * inp5 * /inp4 * /inp3 * inp2 * inp1 * /inp0

The exact syntax depends on the program, once again, but most advanced programs allow abbreviated entry of a similar format.

8.2.2.2   *Truth table entry*

A truth table, as we have seen, defines all the outputs for every active input combination. Each line of the table defines an input condition in terms of HIGH (H or 1), LOW (L or 0) or DON'T CARE (− or X). The outputs may be specified in a similar manner or, in the case of some dedicated PLA or PLS formats, by an 'A' or a '.'. The 'A' indicates that the output is attached to that particular input combination, and will respond if it is applied; the '.' means that the output will not respond to that input pattern.

Some logic compilers will accept design information in truth table format, usually with inputs and outputs defined as '1's and '0's. They may demand that every possible input combination is defined, which would make it unwieldy for more than five or six inputs. However, there would be the advantage that the same data could be used for simulation/test vector definition.

8.2.2.3   *State machine definition*

State machines are specified most clearly as state diagrams. It is possible to convert each state transition into a logic equation, but it is usually more convenient to write each transition in terms of the start and finish states and the required input condition. There are four possible stages involved in converting state diagram information into a format that a logic compiler will recognise.

Firstly the states themselves must be defined. Each state will normally have a name on the state diagram, and will be associated with a particular pattern of HIGHs and LOWs in the state register. The simplest way to encode each state name is to allocate it a number corresponding to the bit pattern of the state register. For example, we could define the nine states of the combinational lock example (Section 2.3.3.2) as:

```
[Q3,Q2,Q1,Q0]
   START = 0000b;
   PASS1 = 0001b;
   PASS2 = 0010b;
   PASS3 = 0011b;
   FAIL1 = 0100b;
   FAIL2 = 0101b;
   FAIL3 = 0110b;
   FAIL4 = 0111b;
  ALARM = 1100b;
```

The first line defines which register elements are referred to in the list; the 'b' defines the numbers as binary, which would be necessary if don't cares were included – these cannot be defined if the common alternatives of octal or hexadecimal numbers are used.

The input conditions can be defined in a similar way, as follows:

```
[I3,I2,I1,I0]
IP8 = 8h;
IP0 = 0h;
IP9 = 9h;
```

This time we have used hexadecimal numbers as these correspond exactly to the inputs we are defining.

The last set of variables is the outputs. We need not define these if the outputs are the same as the state variables, as in a simple Moore machine. Otherwise they may be defined as the state variables or inputs. In our combination lock we had only one output, a signal to unlock the lock; let us suppose that we also have an alarm (AL) which is set when the system has had two invalid numbers entered. The output functions become:

```
[UN,AL]
NULL    = 00b;
UNLOCK = 10b;
AL__OUT = 01b;
```

We are now in a position to define the state transitions. The usual formats are:

WHILE [] ... IF [] ... THEN [] ... WITH [] ... ELSE []

and

STATE [] ... CASE [] [] ... WITH [] ... ELSE [] ... ENDCASE

The arguments of the WHILE and STATE operators are the present state; IF and CASE define the jump conditions, while the next state is the argument of the THEN operator or the second argument of CASE. WITH defines an optional output condition and ELSE the default state if none of the jump conditions is true. We can illustrate this by writing the state equations for the combination lock.

```
WHILE [START]
    IF [IP8] THEN [PASS1] WITH [NULL]
    ELSE [FAIL1] WITH [NULL]

WHILE [PASS1]
    IF [IP0] THEN [PASS2] WITH [NULL]
    ELSE [FAIL2] WITH [NULL]

WHILE [PASS2]
    IF [IP9] THEN [PASS3] WITH [UNLOCK]
    ELSE [FAIL3] WITH [NULL]

WHILE [FAIL1]
    IF [IP8] THEN [FAIL2] WITH [NULL]
    ELSE [ALARM] WITH [AL_OUT]

WHILE [FAIL2]
    IF [IP0] THEN [FAIL3] WITH [NULL]
    ELSE [ALARM] WITH [AL_OUT]

WHILE [FAIL3]
    IF [IP9] THEN [FAIL4] WITH [UNLOCK]
    ELSE [ALARM] WITH [AL_OUT]
```

A logic compiler should generate a state table the same as we produced in Section 5.2.2.3. The 12 transitions defined above would thus be condensed to just five by a process of logic minimisation. This is a feature we will consider later.

8.2.2.4 Schematic capture

Although more suited to work stations or larger computers, some schematic capture programs are available for personal computers. Most PLD compilers make use of general purpose schematic capture programs rather than incorporate them into dedicated PLD programs.

A logic diagram is generated by putting logic symbols onto a screen and interconnecting them as required. The logic symbols may be basic gates and flip-flops or more complex functions, often defined by reference to a standard logic family such as the 74-series. This diagram is converted to a netlist based on primitive logic functions; this forms the interface to the PLD compiler.

The task of the logic compiler is therefore to convert the netlist to logic equations, which can then be compiled to a fuse map in the normal way.

The advantage of schematic capture is that designers who are used to designing general logic systems do not have to learn a new methodology in order to use PLDs. The other side of the coin is that the design is so far removed from the target device that it is difficult to specify the target with any efficiency. Either the target may be more complex than necessary, or if the design will not fit, it may be difficult to see where it may be modified to allow a particular PLD to be used.

One feature of schematic capture packages which does make them appropriate to systems including PLDs is the hierarchical method of specifying logic. Thus, the design can be specified in several layers; the top layer could show the whole system in terms of complex blocks, each block being expanded into a more detailed circuit diagram on lower layers. A PLD could, therefore, be specified by primitive logic functions on the lowest level, and represented by a single symbol on a higher level. This allows PLDs to be integrated into a whole system design.

8.2.3 Logic optimisation

The way in which a designer specifies a logic function may not be the most effective way of implementing it in programmable logic. There are two aspects to this; firstly a designer may inadvertently make his design over-complicated, for example, by unnecessarily duplicating functions. Secondly, he may deliberately choose to specify the logic in terms of intermediate functions, or by using four or five levels of gating, particularly if the design is being converted from a discrete logic solution.

In either case it is possible that the design, as stated, will contain too many logic terms for the target device under consideration. We have seen that Karnaugh maps may be used to reduce the number of logic terms by combining cells in pairs, fours, eights and so on, but this manual technique becomes unwieldy with more than seven or eight inputs. Fortunately, this type of operation can be carried out by a computer, although it is often memory-intensive and time consuming.

This is not the place for detailed descriptions of computerised logic minimisation methods, except to mention the common ones and give an overview of the general process. The old standard was the Quine–McCluskey method, but this has now been largely superseded by the Espresso algorithm. Other optimisation routines also exist, some being proprietary to one compiler, but they should all yield similar results.

A design will often not be written or drawn in basic sum of products format. It is more usual to include intermediate logic functions, or compound functions involving levels of parenthesis. The first step of an optimiser is to expand, or flatten the logic into a basic sum of products format. Any feedback loops will be cut, effectively separating inputs from fed back outputs. This stage is equivalent to filling in the Karnaugh map.

The second stage depends on the target device. For example, a PLA will require that the number of terms is minimised in total, whereas a PAL needs each output minimised separately. Further considerations may be register synthesis, when flip-flops may have to be built from existing gates or set to the optimum type according to the needs of the design, or bringing unused I/Os into play to make up for a shortfall of product terms. Alternatively, in those devices with variable product term numbers (e.g. 22V10), some output functions may need to be restricted to certain pins.

The designer should finish up with a set of equations which fulfils his original logic specification even though they may not look the same. This is fine if the design has been specified correctly, in which case it should fit into the most economical device, but if changes need to be made, it may be difficult to see which part of the design is incorrect. Logic minimisation is thus a powerful tool, but one which should be used with some care.

8.2.4   Device fitting

8.2.4.1   Standard PLDs

We discussed device selection in Chapter 7, but it also has a bearing on the software tools.

Many compilers require that the target device is specified as part of the basic design file. This is only reasonable as logic checking, for example, can only be performed fully if the target device is known. A design with 12 inputs and 9 outputs will be quite acceptable for a 22V10 device, but not for a 20V8.

On the other hand, it might be desirable to know that a certain block of logic will function as desired without committing it to a specific PLD. Some compilers, then, have the option of performing logic checking before device specification. This means that the designer is not constrained by device architecture while he is specifying his logic system, and the logic is specified and verified before he needs to specify the target device.

Compilers which work in this way often have a *device fitter* to complete the logic compilation. The fitter will usually have two functions. Firstly, it will search its device library to find all the devices which have the capacity for the specified logic. Secondly, after the designer has selected from the list of possible targets, it will allocate logic terms to the physical components in the device and generate a fuse map.

8.2.4.2   LSI PLDs

Fitting logic into a standard PLD is relatively straightforward. Each output has well defined logic terms associated with it, and inputs are connected directly into the logic array. This situation does not apply to most LSI PLDs. In FPGAs any cell can be connected to the I/O in a very free manner; the logic itself may not be capable of implementation in a 'normal' sum of products format. The software has to decide how to configure the logic, where each logic element must be placed in the cell array, and how the cells are to be connected to each other and to the I/O cells. This function, called place and route, is usually beyond the scope of standard logic compilers and requires software dedicated to the particular array. The logic compiler may be able to configure the logic into a format suitable for the type of logic cell in a particular FPGA, and produce a list of the logical connections between them. The place and route software will then map these connections onto the physical layout of the cell array.

8.2.5 Design verification

8.2.5.1 Fault grading

It is to be assumed that every gate, or product term, in a logic design has some purpose or, presumably, it should not be there. However, we saw in Chapter 7 that some designs may have circuit nodes which cannot be detected, if faulty, by input vectors. A perfect design should contain none of these *undetectable faults*. In practice, this will not always be the case, at least at the first attempt. Design software should, therefore, be able to tell the designer if any parts of his circuit have hidden faults.

This feature is known as fault grading and is part of the design verification process. Having completed a design, a designer will usually try to simulate it by defining a set of test vectors which force the outputs to measurable states which can be compared with the expected behaviour of the design. Fault grading works by forcing each node in turn to a stuck-at-one fault and a stuck-at-zero fault, and seeing if this makes a detectable difference to the outputs.

The fault grading should report the percentage of detected faults and list those which have not been detected. The designer can then design further tests to increase the number of faults which are detected. Eventually, he may decide that some faults are undetectable. He must then modify the design if he does not want to finish up with a circuit which is only partly testable. It could mean that the logic is not minimised, for that implies that some parts of the logic are redundant, as they could be combined with other logic elements. Minimisation will often remove undetectable faults.

Proper use of fault grading will ensure that the design is fully testable and that the test vectors give 100% fault coverage.

8.2.5.2 Timing simulation

Once a set of test vectors has been established with maximum possible fault coverage, the designer can be reasonably confident that his circuit will work in the intended application – except for one important aspect. The output signals must be established in time to provide the stimulus needed for whatever circuit they are driving. Equally important, input signals must arrive at the PLD with the correct timing to meet setup and hold times, and other internal requirements.

This can be verified if timing simulation is available in the logic compiler package. Provided that the timing of the external signals is known, this information can be added to the test vector information and a timing simulator will predict worst case delays through the PLD, and warn of any potential violations or races which might lead to metastability problems, glitches or the like.

Because accurate internal device modelling is needed for providing this function, timing simulation is more usual in design software generated by the device manufacturer. It is more important in the more complex PLDs where timing through different device paths is not so obvious; often the timing in simple PLDs can be calculated manually from the device data sheet.

8.2.6   Other features

8.2.6.1   Cross assembly

From discussions in earlier chapters, it should be clear that a given logic function can often be programmed into several different devices. Sometimes, for technical or commercial reasons, there may be a change to the basic device used for a particular logic function. It would, of course, be possible to edit the design file to specify a different device. This may not be possible for ethical or logistical reasons, so let us imagine a situation where we have only a JEDEC file, or just a master device.

A cross assembler, possibly with an uploader, gives us the opportunity to convert from one device type to another. The JEDEC file is first converted to a set of logic equations and these are recompiled into a fuse map for the alternate device. Pin names for the reconstituted equations will normally bear no relationship to the original signal names; they are usually christened with appropriate labels such as P1 for pin one, and so on.

8.2.6.2   Merging

One important reason for using PLDs is to reduce the number of packages on a printed circuit board. The ingenuity of device manufacturers does not stand still, though, so today's high complexity device will be superseded by even more complex parts tomorrow. The ability to merge several designs into a single device makes it possible to upgrade painlessly.

The advantages of this are not only economic. Overall performance will probably improve as well. Signal paths will be transferred from the PCB to the chip with a consequent reduction of capacitance and track length. There will probably be a reduction in power consumption also, as one device does the work of many.

When designing FPGAs and other high density PLDs, merging can be a powerful design tool. It will probably be easier to break the design into modules, which can be designed individually and functionally verified, and then merged into a single system.

Chapter 9 includes a selection of logic macros which can be used as the basis of module generation.

8.2.6.3 Documentation

'The job is not finished until the paperwork is done.' Documentation is one of the more irksome tasks of the design engineer, so the more that is done by the computer the happier and more efficient he will be. Typical reports which might be required are:

- Original equations/state diagram
- Optimised equations
- Fuse map
- Chip diagram
- Test vectors
- Resource utilisation

This should be sufficient information for anyone to understand how the design was made and what scope there is for changes, should they be necessary.

8.2.7 Current software

There are two different kinds of design software, manufacturer specific and manufacturer independent. The former is usually issued in support of a range of PLDs, the latter as a commercial venture.

If we look at manufacturer specific products first we can see that these too fall into two camps. Some are issued free of charge, or at least with only a nominal charge, whilst others cost a 'commercial' price. It is arguable as to whether one is justified in charging a relatively high price for software which ties the user to one range of PLDs, but these manufacturers are still in business so some people must think so.

All manufacturer-specific products suffer from the obvious drawback that one is restricted in the number of PLD types which are supported. In order to use other devices, then, another program must be acquired and learned. Not all the features described above are necessarily available; often just logic compilation and basic simulation to provide test vectors are included, although a cross compiler from other manufacturers' devices may well be found!

The situation for independent software is different. These suppliers are in competition with (sometimes) cheap manufacturer software, as well as each other, so every possible feature will usually be included. The chief extra is the range of devices covered, although good relations with the device manufacturers are needed to keep up with new developments.

Full lists of software suppliers are given in Appendix 2. As a conclusion to this section, anybody who is fully committed to PLDs would be best advised to purchase an independent software package; those who might be testing the water should first consider acquiring a program such as AMD's Palasm, which is cheap but still covers a wide range of PLDs.

8.3 PROGRAMMING EQUIPMENT

8.3.1 Options for programming

8.3.1.1 *Build or buy?*

The programmer is the most vital part of the design support equipment; clearly, unless one is able to program a PLD no amount of sophisticated software and design expertise is going to produce a usable device. Most manufacturers publish a programming specification which informs the user which voltages to apply in what sequence and for how long in order to blow each fuse, or load each cell. It is probably not beyond the scope of most competent engineers to build a circuit to implement these instructions. There must also be an interface to accept the data from the design source and translate this into fuse locations. There will probably be the need to accommodate several manufacturers and package styles and, maybe, different technologies.

The programmer is not then such a simple piece of equipment as might appear at first sight. We must also consider the case of programming rejects; most manufacturers accept that their devices are not 100 per cent perfect for, as we have stated before, there is no way to test programmability on non-erasable PLDs. They normally offer to replace programming rejects but need to protect themselves against losses caused by equipment which does not meet their programming specification. Device manufacturers will approve equipment made by specialist companies after extensive qualification tests but, understandably, do not have the resources to test every programmer built by potential PLD users. In the long run then, it is more cost effective and technically better to use a commercial programmer.

In addition to the simple commercial reasons above, dedicated programmer makers are usually in close touch with the device manufacturers and are able to offer programming support to new devices as soon as, or even before, the devices themselves are available. They are also able to take advantage of improvements in programming procedures, or *algorithms*, which may have been introduced to increase yield or reliability.

8.3.1.2 *How complex?*

There is a wide choice of commercial programmers, both in terms of the number of manufacturers and the options available. At the lowest end of the range are programmers which will cater for a small range of devices and offer only a limited number of functions. We have described various families of PLD and some equipment is able to programme just one technology. Some programmers, on the other hand, will cope with any PLD on the market, it is claimed.

In any design situation the programmer must be capable of receiving data, storing it and then using it as the basis for generating the fuse pattern for whatever device is being programmed. The simplest useful system is a programmer with a memory and serial port which allows it to be driven from a terminal or personal computer. This is the minimum configuration which can be considered for designing PLDs. The personal computer is used for the design itself, using a software package as described above, and the fusing information then loaded into the programmer which produces the finished device. Alternatively, the fuse information can be read from a 'master' device which has been created elsewhere, and the data then used to create further copies. The disadvantage of working from a master is that test vectors cannot be used.

More comprehensive, but still tied to a personal computer (PC), are programmers which are driven directly from the PC data bus. They are little more than a set of programmable power supplies and analog switches. The manufacturer's specifications are stored in the computer along with the fusing information. These offer the most versatile form of programmer since they may be updated simply by changing a floppy disk. They are often an economic solution as the intelligence which has to be built into a stand-alone machine is provided by the PC itself. Even if a PC has to be purchased in addition to the programmer, the overall cost is unlikely to be higher than a stand-alone programmer and the PC will be needed for design purposes anyway.

Universal programmers are the most extensive type of machine; they are intended to cover all programmable devices, unlike simple programmers which may have a limited range, and are usually constructed in two parts. The basic chassis contains the programmable power supplies for generating the programming waveforms, together with a keyboard, display screen and memory for storing the fuse pattern. The different ranges of PLD are covered by a series of modules which plug into the chassis. These modules contain the programming specifications, usually in an EPROM, and the circuitry for switching the appropriate waveforms to the socket pins according to the fuse pattern required.

The design data can be entered from the keyboard, or via a serial port from a separate computer. A few sophisticated programmers also have a built-in compiler which can accept logic equations directly, and therefore act as a stand-alone PLD development system. They may have a built-in disk drive to enable designs to be stored for future recall. Universal programmers without the built-in compiler will usually accept data from the keyboard in truth table or fuse chart format; they therefore need design software backup for designers who are not comfortable with those methods of entry.

8.3.2 Production programming

8.3.2.1 *Where to program*

Having successfully completed a design the need arises for production programming. There are three ways in which this can be undertaken; the production/test department of the company can undertake the work, the supplier of the PLDs can program at source, or the programming can be sub-contracted to a specialist programming house. As with most of the situations encountered so far there are arguments for and against each approach.

Keeping the programming in-house means that a better control of stock and schedules can be maintained. If the same type of PLD is used in more than one application then a single purchase can be made and the production planning made more in line with last minute requirements. It should also lead to a wider choice of PLD supplier since one is not restricted to buying from the company that has details of your design. Programming in-house is also likely to be cheaper, provided that quantities are fairly high, because you are paying only for your own labour instead of a profit margin on somebody else's labour.

Against that there is the problem of rejects which, although they are usually replaced, mean ordering extra product and raising extra paperwork to return them. In most cases the production department will need its own programmer so that the design and production can be independent of each other. This will attract an overhead of something in the order of £1000p.a. ($1000p.a.) causing an added cost of 10p (10c) plus labour at a usage rate of 10 000 devices per year. The break-even point compared with an outside programming facility depends on the relative labour charge compared with equipment overhead, which will be much less for the outsider programming larger quantities.

The third option, of using an independent sub-contractor, has many of the advantages of both in-house and third-party programming. It can be particularly beneficial if several programs involving different manufacturers' products are being used. In this case the sub-contractor can provide a mini-kitting service and make the procurement of programmable devices no different from any other component. Costs are usually relatively low because the sub-contractor will be dealing with a high throughput of PLDs.

The argument thus revolves around economics again and will clearly depend on individual circumstances. There is certainly a case for considering buying ready programmed, or subcontracting the programming, unless the volume involved is in the thousands. In that case it may be worth considering moving on from PLDs as we shall discuss later.

8.3.2.2 *How to program*

Assuming that one has made the decision to program in house, the question still remains as to which programmer to use. The main features of programmers have already been described, but there are other aspects which are relevant to the production situation.

Except in the smallest companies, it is not a good idea for designers and production to share a machine. There are bound to be conflicts which will result in delayed production schedules or hold-ups in designs. The production department must be able to provide both the finance and labour for programming.

Apart from reliability and ease of use, which are important to any machine, programming time is crucial if a high throughput is expected. Although parameters such as pulse widths are specified by device manufacturers, there is still a significant difference in the time it takes to program a single device when machines are measured. This will clearly have an impact on the cost of production programming, although it would be of only minor importance in a design lab.

If volumes are really high, it might be worth buying a device handler to speed up throughput. In practice, it takes six to eight seconds for an experienced operator to insert and remove a device from a programmer socket. This time can be cut to one or two seconds by an automatic handler; thus, with a programming time of two seconds, the throughput will be doubled from 450 per hour to 900. Other considerations are operator boredom and handler contact reliability; handlers can mis-program devices much faster than bored operators if there is a contact problem.

One final issue which should be addressed is device identification after programming. This is particularly important if more than one program is used on one base device, or if the security fuses are blown during programming. One option is to use a printed sticky label which can be applied by hand or by an attachment to a device handler. Many different materials and adhesives are available to withstand the various solvents and heat treatments used in PCB assembly.

An alternative is to print directly onto the device with an ink that bonds to the material used to encapsulate the device. Again, machines are available for manual or handler operation.

8.4 DESIGNING OUT PLDs

8.4.1 Problems of scale

So far, we have tended to look at PLDs as the only solution to implementing logic although we have mentioned the other approaches, standard circuits, gate arrays, cell arrays and full custom. In Chapter 7 we looked at the economics of replacing standard circuits with PLDs and found that a replacement factor of 2 to 3, or higher for more complex PLDs, was likely to be economic. A similar sum can be calculated for comparing PLDs with other ASICs (*application specific ICs*). If we do this we have to add another factor to the cost of ownership; that is the start-up or design cost of a masked ASIC.

Even assuming that the customer does most of the work, the mask costs for a gate array are likely to be at least £5000 ($5000), and the designer would need a full CAE system to implement the design. One gate array might replace 5 PLDs, although this is dependent entirely on the type of PLD and size of gate array but, with this assumption, 1000 arrays at £5 ($5) would cost the same as the equivalent circuit in PLDs at £2 ($2). There are other factors to be taken into account; time to sample may be 3–4 weeks for a gate array, mistakes cost another mask charge and a commitment to full production must be made. On the other hand, assembly costs will be lower and performance probably better for the array. Overall there will be a breakpoint of at least 1000 where a masked ASIC becomes the preferred solution.

Although this book is an exhortation to use PLDs as the ideal way of building logic circuits, considering ease of design and economics as the prime motivations, this section will consider PLDs as stepping stones to even more cost-effective solutions. The potential user must perform a cost analysis based on the above arguments to decide which is the optimum route for him.

8.4.2   Hard array logic

Hard array logic, or HAL, was introduced as an answer to the problem outlined above, that masked devices will generally be cheaper than programmable once a threshold volume has been passed. PLDs, as we have seen, need additional circuitry to allow the fuse array to be addressed from the input pins, effectively bypassing the logic circuit. This is the main reason why they are more expensive than their masked equivalents. The other reason for them being less preferable is the fact that they need programming. This is a distinct advantage in the start-up phase of a design; it allows changes to be introduced painlessly, and reduces the risks involved in holding component stocks. Once a product is established in terms of design and run rate, programming can become an expensive and irksome additional process.

The HAL has the same basic structure as its equivalent PAL, but the logic function is defined by a mask instead of by fuses. It is otherwise a complete replacement part in terms of pinning, function and performance. A PAL can be used for the design and early production phase of a product but once the product is established, all, or part, of the production can be taken over by a HAL. Like all custom masked devices, supply quantities of HALs have to be committed several months ahead; if the HAL supply is planned conservatively then any extra product can be met by PALs. This gives the user the best of both worlds; the cost savings of a masked ASIC with the volume flexibility of a programmable device.

8.4.3   Conversion to a gate array

If a product is being designed with the intention, or at least expectation, that it will be produced in high volume, then it is likely that the random logic will be designed into a masked ASIC. The initial steps of the design are the same as for a PLD; that is logic input and design verification. Thus, there is no more work involved in designing the circuit as a PLD than there would be in going straight to a gate array or cell array. There is the advantage that a device can be available within minutes of completing the design, compared with the weeks usually needed for masked samples to appear.

Instant availability of hardware to test is likely to be a benefit to the designer but is there a price to pay? If the design data has to be entered twice then clearly there is, but if the same database can be used for both then there is no problem. Many of the CAE system manufacturers now offer the facility of using standard PLD software files as a design input. The CAE will convert these to gate or cell based primitives for the target array just as it does for standard logic devices. PLDs can therefore be used as a powerful design tool for masked array designers. They also provide hardware back-up in case there is a failure with the masked array.

The design flow, using PLDs as an intermediate state, then appears as in Figure 8.1. In a top-down design the functions intended for the masked device must be partitioned for PLDs. As these are not intended to go into production, a 'minimum cost' solution is not necessary; a 'minimum wiring' solution may even be preferred. The design data for each PLD can then be entered; when all the PLDs have been entered the data may be transferred to the array design and the logic simulated as a total entity. Any changes necessary can be made to the individual blocks and resimulated until the overall function is correct. The design file can then be used for laying out the array, while the PLD files are used for creating devices containing the required logic.

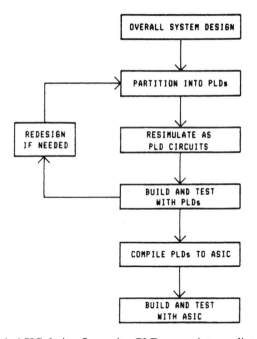

Fig. 8.1 ASIC design flow using PLDs as an intermediate stage.

The system can be built and tested before the array samples are produced, or even before any financial commitment is made to producing masks. It also becomes possible to build prototypes for demonstration or early production, and effectively come to the market quicker by using this approach. Some of the risks involved in using masked ASICs are substantially reduced if PLDs are used as an intermediate stage in the design; the additional cost and work is insignificant in comparison with the benefits obtained.

Chapter 9
Programmable Logic Applications

9.1 MACRO ELEMENTS

9.1.1 Introduction

Applications information can be presented in two ways, each with its own use. One way is to look at the various uses to which devices have been put, the purpose being to trigger ideas in the mind of the designer who may be contemplating similar circuits. The second way is to provide the designer with building bricks, called *macros*, which enable him to synthesise circuits as he might do from standard logic families. This is the approach in this section, the first part of which covers the combinational circuits. Because most designers are familiar with 74 series and 4000 series numbering, these are used as the basis of the macro descriptions.

. Three methods of describing the macros are used, where appropriate; these are logic equations, Karnaugh map and truth table. The macro descriptions can therefore be used by a designer using any of the standard methods of entering logic information into a PLD.

9.1.2 Simple gates

9.1.2.1 AND gates

74 series types; 08, 11, 21
4000 series types: 4073, 4081, 4082

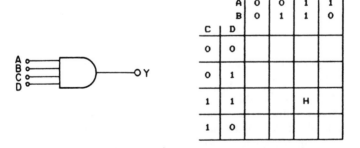

Fig. 9.1 AND symbol and Karnaugh map.

Logic equation: $Y = A * B * C * \dots$
Truth table:

	Active level – H			
A	*B*	*C*	....	*Y*
H	H	H	....	A

9.1.2.2 NAND gates

74 series types: 00, 10, 20, 30, 133
4000 series types: 4011, 4012, 4023, 4068

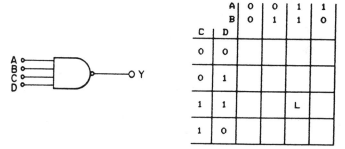

Fig. 9.2 NAND symbol and Karnaugh map.

Logic equation: $/Y = A * B * C* \dots$
Truth table:

	Active level – L			
A	*B*	*C*	....	*Y*
H	H	H	....	A

9.1.2.3 OR gates

74 series type: 32
4000 series types: 4071, 4072, 4075

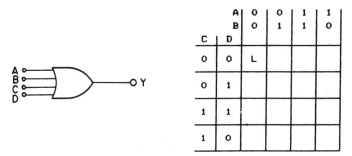

Fig. 9.3 OR symbol and Karnaugh map.

Logic equation: $Y = A + B + C + \ldots$
This may be described in a single AND term as:

$$/Y = /A * /B * /C * \ldots$$

Truth table:

	Active level – L			
A	B	C	....	Y
L	L	L	....	A

9.1.2.4　NOR gates

74 series types: 02, 27
4000 series types: 4000, 4001, 4002, 4025, 4078

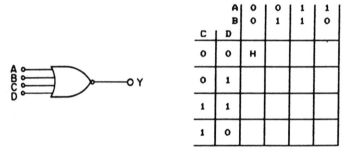

		A B	0 0	0 1	1 1	1 0
C	D					
0	0		H			
0	1					
1	1					
1	0					

Fig. 9.4 NOR symbol and Karnaugh map.

Logic equation: $/Y = A + B + C + \ldots$
This may be described in a single AND term as:

$$Y = /A * /B * /C* \ldots$$

Truth table:

	Active level – H			
A	B	C	....	Y
L	L	L	....	A

9.1.3　Complex gates

9.1.3.1　AND-OR-INVERT gates

74 series types: 51, 54, 64
4000 series types: 4019, 4085, 4086, 4506

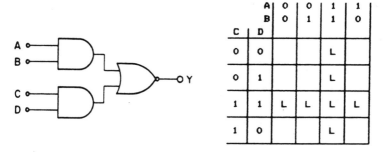

| A | 0 | 0 | 1 | 1 |
| B | 0 | 1 | 1 | 0 |

C	D				
0	0			L	
0	1			L	
1	1	L	L	L	L
1	0			L	

Fig. 9.5 AOI symbol and Karnaugh map.

Logic equation: /Y = A * B + C * D
Truth table:

	Active level – L			
A	*B*	*C*	*D*	*Y*
H	H	–	–	A
–	–	H	H	A

9.1.3.2 Exclusive-OR gates

74 series types: 86, 135
4000 series types: 4030, 4070, 4077, 4507

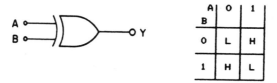

| A | 0 | 1 |
B		
0	L	H
1	H	L

Fig. 9.6 Exclusive-OR symbol and Karnaugh map.

Logic equation: Y = A : + : B
This may be expanded as:
either

$$Y = A * /B + /A * B$$

or

$$/Y = A * B + /A * /B$$

The respective truth tables are:

Active level – H		
A	B	Y
H	L	A
L	H	A

Active level – L		
A	B	Y
H	H	A
L	L	A

9.1.4 Controlled outputs

9.1.4.1 Tri-state buffer or gate

74 series types: 125, 126, 134, 240, 241, 244
4000 series types: 40097, 40098, 4502, 4503

Fig. 9.7 Tri-state buffer symbols.

Tri-state operation is specified in logic equations by means of an extension to the signal name, thus:

Y.ext = E
Y = A

where 'ext' is 'TRST', 'TST', 'OE' or some other abbreviation, depending on which logic compiler is being used. For an active-LOW enable this becomes:

Y.ext = /E
Y = A

The tri-state control terms in a truth table are usually collected in a separate section below the logic terms. In the following table, outputs 3 and 2 have active-HIGH control, outputs 1 and 0 have active-LOW control:

	E	A	B	....	Y3	Y2	Y1	Y0
D3	H	–	–	....				
D2	H	–	–	....				
D1	L	–	–	....				
D0	L	–	–	....				

9.1.4.2 Open collector outputs

74 series types: 01, 03, 26, 33

Open collector outputs, as such, are available on only a few PLEs and very few PLAs. However, a quasi open collector output can be formed with a tri-state output. The condition to be defined is that the output is LOW when the logic is true, and high-impedance when the logic is not true. Because the tri-state must be defined with a single AND term only single product term functions can be given an 'open collector' in this way. The equations for open collector NAND and open collector OR are respectively:

/Y1.TRST = A ∗ B ∗ C ∗ ....
/Y1 = A ∗ B ∗ C ∗ ....

/Y0.TRST = /A ∗ /B ∗ /C ∗ ....
/Y0 = /A ∗ /B ∗ /C ∗ ....

The truth table for NAND (Y1) and OR (Y0) is:

			Active level		
				L	L
A	B	C	....	Y1	Y0
H	H	H	....	A	.
L	L	L	....	.	A
.					
.					
D1	H	H	H	....	⌐
D0	L	L	L	....	

9.1.4.3 Transceivers

74 series types: 242, 243, 245

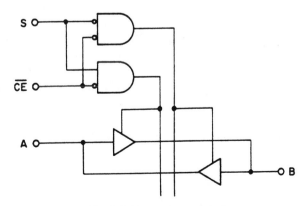

Fig. 9.8 Transceiver symbol.

Logic equations:

B.TRST = /CE * S
B = A
A.TRST = /CE * /S
A = B

To build this function, A and B must be bidirectional pins; AI and BI refer to A and B in the AND array, while AO and BO refer to the OR array in the following truth table:

			Active level			
					H	H
	CE	S	AI	BI	AO	BO
	–	–	H	–	.	A
	–	–	–	H	A	.
	.					
	.					
D1	L	H	–	–	⌐	
D0	L	L	–	–		

9.1.5 Decoders, encoders and multiplexers

9.1.5.1 Decoders

74 series types: 42, 138, 139, 154, 155, 156
4000 series types: 4514, 4515, 4528, 4555, 4556

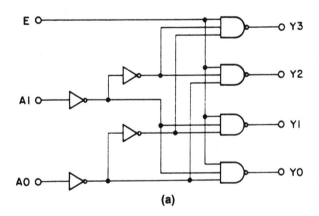

(a)

Fig. 9.9 1-of-4 Decoder circuit and Karnaugh maps.

Y3

A1	0	0	1	1
A0	0	1	1	0
E				
0				
1			L	

Y2

A1	0	0	1	1
A0	0	1	1	0
E				
0				
1				L

YI

A1	0	0	1	1
A0	0	1	1	0
E				
0				
1		L		

Y0

A1	0	0	1	1
A0	0	1	1	0
E				
0				
1	L			

(b)

Fig. 9.9 (*cont.*)

Logic equations:

$$/Y0 = E * /A1 * /A0$$
$$/Y1 = E * /A1 * A0$$
$$/Y2 = E * A1 * /A0$$
$$/Y3 = E * A1 * A0$$

Truth table:

	Active level		L	L	L	L
E	A1	A0	Y3	Y2	Y1	Y0
H	L	L	.	.	.	A
H	L	H	.	.	A	.
H	H	L	.	A	.	.
H	H	H	A	.	.	.

9.1.5.2 Priority encoders

74 series types: 147, 148
4000 series types: 40147, 4532

I3	I2	I1	I0	A1	A0
L	X	X	X	H	H
H	L	X	X	H	L
H	H	L	X	L	H
H	H	H	L	L	L

A1

I1 \\ I0 \ I3 I2	0 0	0 1	1 1	1 0
0 0			L	
0 1			L	
1 1				
1 0			L	

A0

I1 \\ I0 \ I3 I2	0 0	0 1	1 1	1 0
0 0				L
0 1				L
1 1				L
1 0			L	L

Fig. 9.10 4-input priority encoder function table and Karnaugh maps.

Logic equations (by reference to the Karnaugh maps we can implement a glitch-free design by overlapping the AND terms):

$$/A1 = I3 * I2 * /I1 + I3 * I2 * /I0$$
$$/A0 = I3 * /I2 + I3 * I1 * /I0$$

Truth table:

I3	I2	I1	I0	L A1	L A0
	Active level				
H	H	L	–	A	.
H	H	–	L	A	.
H	L	–	–	.	A
H	–	H	L	.	A

9.1.5.3 Multiplexers

74 series types: 150,151, 153, 157, 158
4000 series types: 4019, 40257, 4539

S1	S0	I3	I2	I1	I0	Y
L	L	X	X	X	H	H
L	L	X	X	X	L	L
L	H	X	X	H	X	H
L	H	X	X	L	X	L
H	L	X	H	X	X	H
H	L	X	L	X	X	L
H	H	H	X	X	X	H
H	H	L	X	X	X	L

Fig. 9.11 4-input multiplexer function table and Karnaugh map.

I2	I1	I0	S1=0 S0=0 I3=0	0 0 1	0 1 1	0 1 0	1 1 0	1 1 1	1 0 1	1 0 0
0	0	0						H		
0	0	1	H	H				H		
0	1	1	H	H	H	H		H		
0	1	0			H	H		H		
1	1	0			H	H		H	H	H
1	1	1	H	H	H	H		H	H	H
1	0	1	H	H				H	H	H
1	0	0						H	H	H

Fig. 9.11 (*cont.*)

Logic equation (referring to the Karnaugh map shows that the minimum solution of four AND terms does not produce any overlapping, so this solution is prone to hazards):

$$Y = S1 * S0 * I3$$
$$+ S1 * /S0 * I2$$
$$+ /S1 * S0 * I1$$
$$+ /S1 * /S0 * I0$$

Truth table:

				Active level H		
S1	S0	I3	I2	I1	I0	Y
H	H	H	–	–	–	A
H	L	–	H	–	–	A
L	H	–	–	H	–	A
L	L	–	–	–	H	A

9.1.6   Arithmetic circuits

9.1.6.1   *Magnitude comparator*

74 series type: 85
4000 series types: 4063, 40085, 4585

A2	A1	A0	B2	B1	B0		A>B	A=B	A<B
H	X	X	L	X	X		H	L	L
H	H	X	H	L	X		H	L	L
L	H	X	L	L	X		H	L	L
H	H	H	H	H	L		H	L	L
H	L	H	H	L	L		H	L	L
L	H	H	L	H	L		H	L	L
L	L	H	L	L	L		H	L	L
H	H	H	H	H	H		L	H	L
H	H	L	H	H	L		L	H	L
H	L	H	H	L	H		L	H	L
etc.									
L	X	X	H	X	X		L	L	H
H	L	X	H	H	X		L	L	H
L	L	X	L	H	X		L	L	H
etc.									

		A2	0	0	0	0	1	1	1	1
		A1	0	0	1	1	1	1	0	0
		A0	0	1	1	0	0	1	1	0
B2	B1	B0								
0	0	0	E	G	G	G	G	G	G	G
0	0	1	S	E	G	G	G	G	G	G
0	1	1	S	S	E	S	G	G	G	G
0	1	0	S	S	G	E	G	G	G	G
1	1	0	S	S	S	S	E	G	S	S
1	1	1	S	S	S	S	S	E	S	S
1	0	1	S	S	S	S	G	G	E	S
1	0	0	S	S	S	S	G	G	G	E

Fig. 9.12 Magnitude comparator function table and Karnaugh map.

Logic equations:
The Karnaugh map is drawn in terms of A>B (G), A=B (E) and A<B (S), from which it may be seen that 'E' is just /G * /S, and that 'G' and 'S' are interchangeable by exchanging 'A' and 'B'. We will just write the equation for the 'G'-terms, as:

$$A > B = /A2 * A0 * /B2 * /B1 * /B0$$
$$+ A1 * /B2 * /B1$$
$$+ A2 * /B2$$
$$+ A2 * A1 * /B1$$
$$+ A2 * A0 * /B1 * /B0$$
$$+ A2 * A1 * A0 * B1 * /B0$$
$$+ /A2 * A1 * A0 * /B2 * /B0$$

Truth table:

| | | | | | | Inputs | | Outputs | | |
| | | | | | | *Bidirectional pins* | | | | |
A2	A1	A0	B2	B1	B0	A > B Active	A < B level	A > B H	A < B H	A = B H
L	–	H	L	L	L	–	–	A	.	.
–	H	–	L	L	–	–	–	A	.	.
H	–	–	L	–	–	–	–	A	.	.
H	H	–	–	L	–	–	–	A	.	.
H	–	H	–	L	L	–	–	A	.	.
H	H	H	–	H	L	–	–	A	.	.
L	H	H	L	–	L	–	–	A	.	.
–	–	–	–	–	–	L	L	.	.	A

9.1.6.2 *Parity generator*

74 series types: 180, 280
4000 series types: 40101, 4531

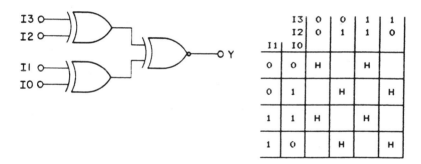

Fig. 9.13 4-input parity generator circuit and Karnaugh map.

Logic equation:

From the Karnaugh map a separate AND term is required for each 'H'-cell, that is eight altogether, and this total doubles for each additional input. A less AND-term intensive design which uses more outputs is to construct exclusive-OR gates and interconnect these. This structure uses up outputs except in those devices which possess internal feedback paths. The equation, for even parity, is:

$$PE = (I3 : + : I2) : + : (I1 : + : I0)$$

The brackets are optional, and are included merely for clarity.

Truth table:

				Bidirectional pins				
				Inputs		Outputs		
$I3$	$I2$	$I1$	$I0$	$X1$	$X0$	$X1$	$X0$	PE
				Active	level	H	H	H
H	H	–	–	–	–	A	.	.
L	L	–	–	–	–	A	.	.
–	–	H	H	–	–	.	A	.
–	–	L	L	–	–	.	A	.
–	–	–	–	H	H	.	.	A
–	–	–	–	L	L	.	.	A

9.1.6.3 Full adder

74 series types: 83, 283
4000 series types: 4008, 4568(BCD)

Fig. 9.14 2-bit adder Karnaugh maps.

	C	0	0	0	0	1	1	1	1
	A1	0	0	1	1	1	1	0	0
	A0	0	1	1	0	0	1	1	0
B1	**B0**								
0	0						H		
0	1			H		H	H		
1	1		H	H	H	H	H	H	H
1	0				H	H	H	H	H

C

Fig. 9.14 (*cont.*)

Analysis of the Karnaugh maps shows that, even with AND term sharing, as in a PLA structure, 16 AND terms are needed to implement this function, which is only half as complex as a standard logic device. In the standard logic families much use is made of exclusive-OR gates, which, as we saw with the parity generator, need to use a bidirectional pin and are thus of limited use in PLDs. It is better to use PROMs, or PLEs, to construct arithmetic functions in PLDs. It is usually possible to generate the address/data code for a PLE with a computer for arithmetic functions; as an example we will list these for the 2-bit adder. The pin/function assignment is:

Cin – A4 $Cout$ – Q2
B1 – A3   S1 – Q1
B0 – A2   S0 – Q0
A1 – A1
A0 – A0

The address/data table, in hexadecimal, is:

	0	*1*	*2*	*3*	*4*	*5*	*6*	*7*	*8*	*9*	*A*	*B*	*C*	*D*	*E*	*F*
0000	0	1	2	3	1	2	3	4	2	3	4	5	3	4	5	6
0010	1	2	3	4	2	3	4	5	3	4	5	6	4	5	6	7

9.1.7 Latches

9.1.7.1 D-Latch

74 series types: 75, 118, 373
4000 series types: 4042, 40373, 4532

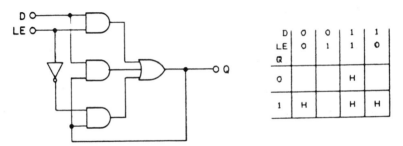

Fig. 9.15 D-latch circuit and Karnaugh map.

Logic equation:

$$Q = D * LE + Q * /LE + D * Q$$

Truth table:

		Bid. Pins	
		I/P	O/P
D	LE	Q	Q
	Active	level	H
H	H	–	A
–	L	H	A
H	–	H	A

9.1.7.2 R–S latches

74 series type: 279
4000 series types: 4043, 4044

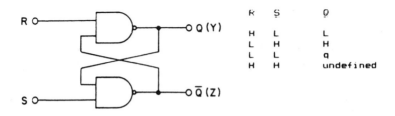

Fig. 9.16 R-S latch circuit and function table.

Logic equations:

$$Q = /(R * /Q)$$
$$/Q = /(S * Q)$$

There may be some confusion here in that Q and /Q are *different* outputs, and not merely complemented versions of one output. In most CAD systems Q and /Q will have to be given different symbols, e.g. Y and Z, and the equations then become:

$$/Y = R * Z$$
$$/Z = S * Y$$

Truth table:

			Bidirectional pins		
	Inputs			Outputs	
R	S	Y	Z	Y	Z
			Active level	H	H
H	–	–	H	A	.
–	H	H	–	.	A

9.1.7.3 Addressable latch

74 series types: 256, 259
4000 series types: 4099, 4724

E	D	A1	A0	Q̄3	Q̄2	Q̄1	Q̄0
L	x	x	x	q3	q2	q1	q0
H	H	H	H	H	q2	q1	q0
H	L	H	H	L	q2	q1	q0
H	H	H	L	q3	H	q1	q0
H	L	H	L	q3	L	q1	q0
H	H	L	H	q3	q2	H	q0
H	L	L	H	q3	q2	L	q0
H	H	L	L	q3	q2	q1	H
H	L	L	L	q3	q2	q1	L

	E	0	0	0	0	1	1	1	1
	D	0	0	1	1	1	1	0	0
	Q̄3	0	1	1	0	0	1	1	0
A1	**A0**								
0	0		H	H			H	H	
0	1		H	H			H	H	
1	1		H	H		H	H		
1	0		H	H			H	H	

Q3

Fig. 9.17 4-bit addressable latch function table and Karnaugh map (Q3 only).

Logic equation (Q3 only):

Q3 = A1 * A0 * E * D
　　+ /E * Q3
　　+ /A1 * E * Q3
　　+ /A0 * E * Q3
　　+ D * Q3 (deglitch term)

Truth table:

A1	A0	E	D	I/P Q3 Active level	O/P Q3 H
H	H	H	H	–	A
–	–	L	–	H	A
L	–	H	–	H	A
–	L	H	–	H	A
–	–	–	H	H	A

9.1.8　Flip-flops

9.1.8.1　D-type flip-flop

74 series types: 74, 173, 174, 175, 273, 374
4000 series types: 4013, 40174, 40175, 40374

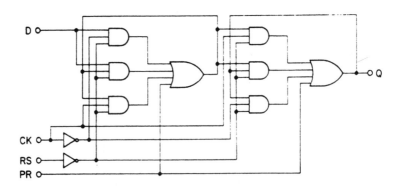

Fig. 9.18 D-type flip-flop circuit diagram.

Logic equations:
　　X = /CK * D * /RS
　　　+ X * D * /RS
　　　+ CK * X * /RS
　　　+ PR

Q=CK * X * /RS
 +Q * X * /RS
 +/CK * Q * /RS
 +PR

Truth table:

				Inputs		Outputs	
						Bidirectional pins	
D	CK	PR	RS	X	Q	X	Q
				Active	*level*	*H*	*H*
H	L	–	L	–	–	A	.
H	–	–	L	H	–	A	.
–	H	–	L	H	–	A	.
–	–	H	–	–	–	A	A
–	H	–	L	H	–	.	A
–	–	–	L	H	H	.	A
–	L	–	L	–	H	.	A

9.1.8.2 J–K flip-flop

74 series types: 107, 109, 112, 113
4000 series types: 4027, 4095, 4096

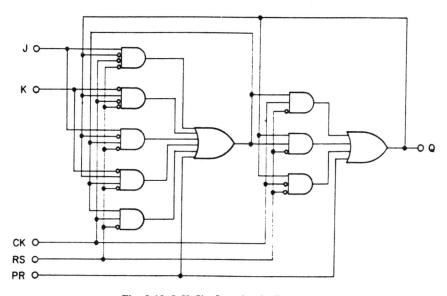

Fig. 9.19 J–K flip-flop circuit diagram.

Logic equations:

$$X = /CK * J * Q * /RS$$
$$+ /CK * /K * /Q * /RS$$
$$+ CK * X * /RS$$
$$+ J * Q * X * /RS$$
$$+ /K * /Q * X * /RS$$
$$+ PR$$
$$Q = CK * X * /RS$$
$$+ X * Q * /RS$$
$$+ /CK * Q * /RS$$
$$+ PR$$

Truth table:

					Inputs		Bidirectional pins Outputs	
J	K	CK	PR	RS	X Active	Q level	X H	Q H
H	–	L	–	L	–	H	A	.
–	L	L	–	L	–	L	A	.
–	–	H	–	L	H	–	A	A
H	–	–	–	L	H	H	A	.
–	L	–	–	L	H	L	A	.
–	–	–	H	–	–	–	A	A
–	–	–	–	L	H	H	.	A
–	–	L	–	L	–	H	.	A

9.2 SEQUENTIAL MACRO ELEMENTS

9.2.1 Introduction

9.2.1.1 *Standard registered circuits*

Examining the list of standard TTL and CMOS functions reveals that the only standard functions using flip-flops are registers and counters. However some of the functions which we have described as combinational can usefully be implemented in registered PLDs. One reason for doing this is to synchronise the output with the system clock, in which case the function may be registered by substituting the equivalent registered device for the combinational device; for example, using a PAL16R8 instead of a PAL16L8.

On the other hand, many of the registered parts have a more complex structure than the combinational devices. For example, the PAL20X4 with its exclusive-OR gates or the PLSs with their complement term may be more efficient at containing arithmetic type functions than the basic AND–OR arrays. We will, therefore, look at some of these structures again to see how they fit into the more complex parts.

9.2.2 Registered combinational functions

9.2.2.1 Addressable register (X-PAL)

A1	A0	E 0 D 0 Q3 0	0 0 1	0 1 1	0 1 0	1 1 0	1 1 1	1 0 1	1 0 0
0	0		H	H			H	H	
0	1		H	H			H	H	
1	1		H	H		H	H		
1	0		H	H			H	H	

Fig. 9.20 4-bit addressable register Karnaugh maps.

Logic equations:

$$/Q3 := A1 * A0 * /D * Q3 * E + A1 * A0 * D * /Q3 * E$$
$$: + : /Q3$$
$$/Q2 := A1 * /A0 * /D * Q2 * E + A1 * /A0 * D * /Q2 * E$$
$$: + : /Q2$$
$$/Q1 := /A1 * A0 * /D * Q1 * E + /A1 * A0 * D * /Q1 * E$$
$$: + : /Q1$$
$$/Q0 := /A1 * /A0 * /D * Q0 * E + /A1 * /A0 * D * /Q0 * E$$
$$: + : /Q0$$

9.2.2.2 Addressable register (PLS)

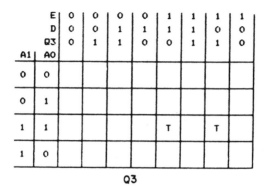

A1	A0	E 0 D 0 Q3 0	0 0 1	0 1 1	0 1 0	1 1 0	1 1 1	1 0 1	1 0 0
0	0								
0	1								
1	1					T		T	
1	0								

Q3

Fig. 9.21 4-bit addressable register Karnaugh maps (J–K flip-flops).

Logic equations:

$$Q3.J = E * A1 * A0 * D * /Q3 + E * A1 * A0 * /D * Q3$$
$$Q3.K = E * A1 * A0 * D * /Q3 + E * A1 * A0 * /D * Q3$$
$$Q2.J = E * A1 * /A0 * D * /Q2 + E * A1 * /A0 * /D * Q2$$
$$Q2.K = E * A1 * /A0 * D * /Q2 + E * A1 * /A0 * /D * Q2$$
$$Q1.J = E * /A1 * A0 * D * /Q1 + E * /A1 * A0 * /D * Q1$$
$$Q1.K = E * /A1 * A0 * D * /Q1 + E * /A1 * A0 * /D * Q1$$
$$Q0.J = E * /A1 * /A0 * D * /Q0 + E * /A1 * /A0 * /D * Q0$$
$$Q0.K = E * /A1 * /A0 * D * /Q0 + E * /A1 * /A0 * /D * Q0$$

Truth table:

Inputs				Present state				Next state			
A1	A0	D	E	Q3	Q2	Q1	Q0	Q3	Q2	Q1	Q0
H	H	H	H	L	–	–	–	0	–	–	–
H	H	L	H	H	–	–	–	0	–	–	–
H	L	H	H	–	L	–	–	–	0	–	–
H	L	L	H	–	H	–	–	–	0	–	–
L	H	H	H	–	–	L	–	–	–	0	–
L	H	L	H	–	–	H	–	–	–	0	–
L	L	H	H	–	–	–	L	–	–	–	0
L	L	L	H	–	–	–	H	–	–	–	0

9.2.2.3 *Registered magnitude comparator*

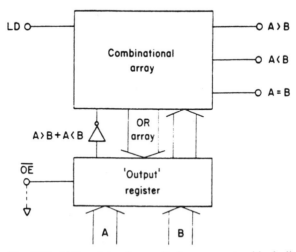

Fig. 9.22 4-bit registered magnitude comparator block diagram.

Description:
The logic equations for this function were detailed in Section 9.1.6.1. By using
a PLS159, the inputs can be loaded into a register and the result of comparison

obtained from the combinational outputs. Using the complement term to feed back A > B and A < B to generate A = B saves an output, leaving three inputs and one output free for extra logic functions.

Truth table:

	I/P		*Present state*								*Output*		
Term	*CT*	*LD*	*A3*	*A2*	*A1*	*A0*	*B3*	*B2*	*B1*	*B0*	*A > B*	*A < B*	*A = B*
00	A	–	H	–	–	–	L	–	–	–	A	.	.
01	A	–	–	H	–	–	L	L	–	–	A	.	.
.													
.													
12	A	–	H	L	–	H	H	L	L	L	A	.	.
13	A	–	L	–	–	–	H	–	–	–	.	A	.
14	A	–	L	L	–	–	–	H	–	–	.	A	.
.													
.													
25	A	–	H	L	L	L	H	L	–	H	.	A	.
26	.	–	–	–	–	–	–	–	–	–	.	.	A
.													
.													
LB	–	H	–	–	–	–	–	–	–	–			
LA	–	H	–	–	–	–	–	–	–	–			
D3	–	–	–	–	–	–	–	–	–	–			
D2	–	–	–	–	–	–	–	–	–	–			
D1	–	–	–	–	–	–	–	–	–	–			

Note: Only part of the truth table is listed as A > B and A < B each require 13 terms. These may be derived from the Karnaugh map in Section 9.1.6.1.

9.2.3 Registers

9.2.3.1 Universal shift register

74 series types: 91, 94, 95, 96, 164, 165, 166, 194, 195, 199, 295, 395
4000 series types: 4014, 4015, 4021, 4034, 4035, 4094, 40104

Logic equations:
$$QA := /S * A \text{ (load)}$$
$$+ R * S * A \text{ (shift right)}$$
$$+ /R * S * QB \text{ (shift left)}$$
$$QB := /S * B$$
$$+ R * S * QA$$
$$+ /R * S * QC$$
$$QC := /S * C$$
$$+ R * S * QB$$
$$+ /R * S * QD$$

$$QD: = /S * D$$
$$+ R * S * QC$$
$$+ /R * S * D$$

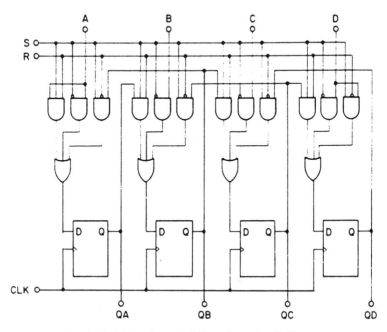

Fig. 9.23 4-bit universal shift register circuit diagram.

Truth table (D-type flip-flops):

| | | | *Inputs* | | | | | *Present state* | | | | *Next state* | | |
R	S	A	B	C	D	QA	QB	QC	QD	QA	QB	QC	QD
–	L	H	–	–	–	–	–	–	–	H	–	–	–
H	H	H	–	–	–	–	–	–	–	H	–	–	–
L	H	–	–	–	–	–	H	–	–	H	–	–	–
–	L	–	H	–	–	–	–	–	–	–	H	–	–
H	H	–	–	–	–	H	–	–	–	–	H	–	–
L	H	–	–	–	–	–	–	H	–	–	H	–	–
–	L	–	–	H	–	–	–	–	–	–	–	H	–
H	H	–	–	–	–	–	H	–	–	–	–	H	–
L	H	–	–	–	–	–	–	–	H	–	–	H	–
–	L	–	–	–	H	–	–	–	–	–	–	–	H
H	H	–	–	–	–	–	–	H	–	–	–	–	H
L	H	–	–	–	H	–	–	–	–	–	–	–	H

9.2.3.2 Registered Barrel Shifter

74 series type: 350 (similar, but unregistered)

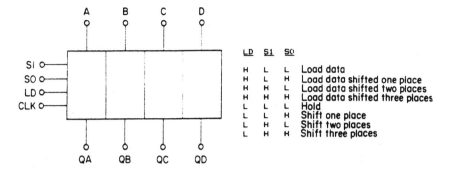

Fig. 9.24 4-bit barrel shifter block diagram and function table.

Logic equations:

$$/QA: = LD * /A * /S1 * /S0$$
$$+ LD * /D * /S1 * S0$$
$$+ LD * /C * S1 * /S0$$
$$+ LD * /B * S1 * S0$$
$$+ /LD * /QA * /S1 * /S0$$
$$+ /LD * /QD * /S1 * S0$$
$$+ /LD * /QC * S1 * /S0$$
$$+ /LD * /QB * S1 * S0$$

The equations for QB, QC and QD follow a similar pattern.

Truth table (D-type flip-flops):

	Inputs						Present state				Next state			
LD	S1	S0	A	B	C	D	QA	QB	QC	QD	QA	QB	QC	QD
H	L	L	L	–	–	–	–	–	–	–	A	.	.	.
H	L	H	–	–	–	L	–	–	–	–	A	.	.	.
H	H	L	–	–	L	–	–	–	–	–	A	.	.	.
H	H	H	–	L	–	–	–	–	–	–	A	.	.	.
L	L	L	–	–	–	–	H	–	–	–	A	.	.	.
L	L	H	–	–	–	–	–	–	–	H	A	.	.	.
L	H	L	–	–	–	–	–	–	H	–	A	.	.	.
L	H	H	–	–	–	–	–	H	–	–	A	.	.	.
H	L	L	–	L	–	–	–	–	–	–	.	A	.	.
etc.														

9.2.4 Counters

9.2.4.1 Ripple counter

74 series types: 90, 92, 93, 197, 290, 293, 390, 393, 490

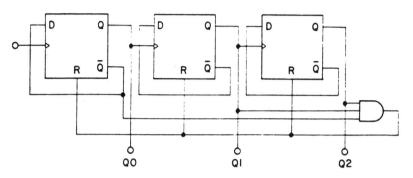

Fig. 9.25 Divide-by-5 ripple counter circuit diagram.

Truth table:

				Bidirectional pins								
			Inputs						*Outputs*			
CK	X0	Q0	X1	Q1	X2	Q2	X0	Q0	X1	Q1	X2	Q2
					Active level		L	L	L	L	L	L
L	–	L	–	–	–	–	A	A	.	.	.	.
–	H	L	–	–	–	–	A	.	.	.	.	.
H	H	–	–	–	–	–	A	.	.	.	.	.
H	L	–	–	–	–	–	.	A	.	.	.	.
–	L	L	–	–	–	–	.	A	.	.	.	.
–	–	L	–	L	–	–	.	.	A	A	.	.
–	–	–	H	L	–	–	.	.	A	.	.	.
.												
.												
.												
–	–	L	–	H	–	H	A	.	A	.	A	.
–	–	H	–	–	–	–	.	A	.	A	.	A
–	–	–	–	L	–	–	.	A	.	A	.	A
–	–	–	–	–	–	L	.	A	.	A	.	A

Logic equations:

$$X0 = /CK * /Q0$$
$$+ X0 * /Q0$$
$$+ X0 * CK$$
$$+ /Q0 * Q1 * Q2 \text{ (reset at count '6')}$$
$$/Q0 = CK * /X0$$
$$+ /Q0 * /X0$$
$$+ /CK * /Q0$$
$$+ Q0 + /Q1 + /Q2 \text{ (reset)}$$

Q1 and Q2 follow the same equations with suffix '0' replaced by '1' and '2', and CK by Q0 and Q1 respectively.

If a different count is required, all that is necessary is to modify the reset term; no reset is required if the count is divide-by-$2n$. A better solution would be to use an asynchronous PLD, although this example shows how flip-flops made from combinational PLDs can be strung together. Asynchronous PLD equations would take the form:

$$Q0 := /Q0$$
$$Q0.CLK = CLK$$

$$Q1 := /Q1$$
$$Q1.CLK = Q0$$
and so on.

A synchronous reset for each input at the count limit (n) would then make the 'divide-by-n' operative.

9.2.4.2 Synchronous counters

74 series types: 160, 161, 162, 163, 168, 169, 190, 191, 192, 193, 568, 569
4000 series types: 4029, 40102, 40103, 40110, 4510, 4516, 4518, 4520, 4521, 4522, 4526

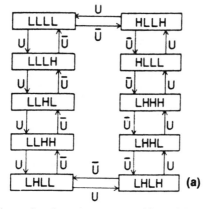

Fig. 9.26 Up/down decade counter state table and Karnaugh maps.

U	0	0	0	0	1	1	1	1
Q3	0	0	1	1	1	1	0	0
Q2	0	1	1	0	0	1	1	0

Q1	Q0								
0	0	H				H			
0	1					H			
1	1							H	
1	0								

Q3

U	0	0	0	0	1	1	1	1
Q3	0	0	1	1	1	1	0	0
Q2	0	1	1	0	0	1	1 .	0

Q1	Q0								
0	0					H		H	
0	1		H					H	
1	1		H						H
1	0		H					H	

Q2

U	0	0	0	0	1	1	1	1
Q3	0	0	1	1	1	1	0	0
Q2	0	1	1	0	0	1	1	0

Q1	Q0								
0	0		H		H				
0	1							H	H
1	1	H	H						
1	0							H	H

Q1

U	0	0	0	0	1	1	1	1
Q3	0	0	1	1	1	1	0	0
Q2	0	1	1	0	0	1	1	0

Q1	Q0								
0	0	H	H		H	H		H	H
0	1								
1	1								
1	0	H	H					H	H

Q0

(b)

Fig. 9.26 (*cont.*)

Logic equations (decade counter):
(D-type flip-flops with inverting outputs are assumed)

$$/Q3 := /U * /Q3 * Q1$$
$$+ /U * /Q3 * Q0$$
$$+ /U * /Q3 * Q2$$
$$+ /U * Q3 * /Q2 * /Q1 * /Q0$$
$$+ U * Q3 * /Q2 * /Q1 * Q0$$
$$+ U * /Q3 * /Q1$$
$$+ U * /Q3 * /Q0$$
$$+ U * /Q3 * /Q2$$

$$/Q2 := /U * /Q3 * /Q2$$
$$+ /U * /Q3 * /Q1 * /Q0$$
$$+ Q3 * /Q2 * /Q1 * Q0$$
$$+ U * /Q2 * /Q1$$
$$+ /Q3 * /Q2 * /Q0$$

$$/Q1 := /U * /Q3 * /Q2 * /Q1$$
$$+ /U * /Q3 * /Q1 * Q0$$
$$+ /U * /Q3 * Q1 * /Q0$$
$$+ Q3 * /Q2 * /Q1 * Q0$$
$$+ U * Q3 * /Q2 * /Q1$$
$$+ U * /Q3 * /Q1 * /Q0$$
$$+ U * /Q3 * Q1 * Q0$$

$$/Q0 := /Q3 * Q0$$
$$+ Q3 * /Q2 * /Q1 * Q0$$

Alternatively, the equations for J–K flip-flops are:

$$Q3.J = /U * /Q2 * /Q1 * /Q0$$
$$+ /U * Q3 * /Q2 * /Q1 * /Q0$$
$$+ U * /Q3 * Q2 * Q1 * Q0$$

$$Q2.J = /U * /Q3 * Q2 * /Q1 * /Q0$$
$$+ /U * Q3 * /Q2 * /Q1 * /Q0$$
$$+ U * /Q3 * Q1 * Q0$$

$$Q1.J = /U * /Q3 * Q2 * /Q1 * /Q0$$
$$+ /U * Q3 * /Q2 * /Q1 * /Q0$$
$$+ /U * /Q3 * Q1 * /Q0$$
$$+ U * /Q3 * Q0$$

$$Q0.J = /Q3$$
$$= Q3 * /Q2 * /Q1$$

The equations for $Q n.K$ are identical to the $Q n.J$ equations. Alternatively, the $Q n.T$ format can be used with those compilers which allow definition of a toggle function.

Truth table:

U	Present state				Next state			
	/Q3	/Q2	/Q1	/Q0	/Q3	/Q2	/Q1	/Q0
L	H	H	H	H	0	–	–	–
L	L	H	H	H	0	0	0	–
H	L	H	H	L	0	–	–	–
H	H	L	L	L	0	–	–	–
H	H	L	H	H	–	0	0	–
H	H	–	L	L	–	0	–	–
L	H	–	L	H	–	–	0	–
H	H	–	–	L	–	–	0	–
–	H	–	–	–	–	–	–	0
–	L	H	H	–	–	–	–	0

9.2.4.3 *Johnson counters*

4000 series types: 4017, 4022

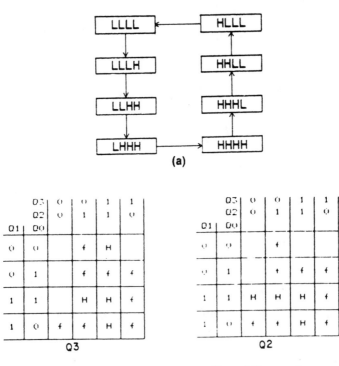

Fig. 9.27 4-bit Johnson counter state table and Karnaugh maps.

Q3→	0	0	1	1
Q2→	0	1	1	0
Q1 Q0				
0 0			f	
0 1	H	f	f	f
1 1	H	H	H	f
1 0	f	f		f

Q1

Q3→	0	0	1	1
Q2→	0	1	1	0
Q1 Q0				
0 0	H	f		
0 1	H	f	f	f
1 1	H	H		f
1 0	f	f		f

Q0

(b)

Fig. 9.27 (*cont.*)

Logic equations:
In the above Karnaugh maps forbidden states have been shown by an 'f'. By including these as an option the equations can be simplified (to those of a shift register). This will work if the circuit starts in a legal state so, if the power-up condition is not known, a jump to a legal state (HHHH) must be included from all the forbidden states. The equations are thus:

$$/Q3 := /Q2$$
$$+Q2 * /Q1 * Q0$$
$$+/Q3 * Q2 * /Q0$$
$$/Q2 := /Q1$$
$$+/Q3 * Q1 * /Q0$$
$$+Q3 * /Q2 * Q1$$
$$/Q1 := /Q0$$
$$+Q2 * /Q1 * Q0$$
$$+Q3 * /Q2 * Q0$$
$$/Q0 := Q3$$
$$+/Q3 * Q2 * /Q1$$
$$+/Q3 * Q1 * /Q0$$

Truth table:
Thanks to the PLS structure, a more efficient solution may be obtained by including all the jumps from forbidden states together, thus:

Present state				Next state			
/Q3	/Q2	/Q1	/Q0	/Q3	/Q2	/Q1	/Q0
H	L	H	–	H	H	H	H
L	–	H	L	H	H	H	H
L	H	L	–	H	H	H	H
H	–	L	H	H	H	H	H
–	H	–	–	H	–	–	–
–	–	H	–	–	H	–	–
–	–	–	H	–	–	H	–
L	–	–	–	–	–	–	H

9.3 MISCELLANEOUS APPLICATIONS

9.3.1 Introduction

This section is devoted to designs which may be incorporated into PLD circuits, but which are not generally available as discrete logic circuits in the 74 series or 4000 series. In some cases they may be available as dedicated LSI circuits, although they are mostly too specialised or not complex enough to warrant an independent existence in a manufacturer's catalogue. Even if they do not fit exactly into a design which you want, they are still worth looking at as they may provide ideas or show how various functions can be implemented in PLDs. In general we list only the logic equations or state table, whichever is more appropriate to the application being described, and the target device.

9.3.2 Specific examples

9.3.2.1 *Waveform generator*

In many applications it is necessary to generate a complex waveform, or several related waveforms. Examples are:

- processor control
- video controllers
- bubble memory drivers

The waveforms are created from a clock and, sometimes, a synchronisation signal. Figure 9.28 shows a block diagram of the basic system which contains two elements, a counter and a decoder. The waveform period will usually be a fixed number of clock periods, so the counter cycles continuously dividing by this number, or is reset by the synchronising pulse at the start of each cycle. The decoder defines the signal level for each number of the count.

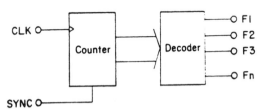

Fig. 9.28 Block diagram of general purpose waveform generator.

Almost any PLD, or combination of PLDs, may be used in this application, depending on the number and complexity of the waveforms. For example, the counter could be a divide-by-4096 (10-bits with a PAL20X10) and the decoder a PLE10P4. This would enable four very complex waveforms to be produced. On the other hand, a PLS155 could produce eight outputs, with a much simpler relationship, but incorporating the counter and decoder in the same device.

As a concrete example let us look at a simple set of signals which might be needed in a video controller. We will see how to produce a frame sync, line sync and picture enable, as shown in Figure 9.29. Frame sync is a half clock frequency signal on lines 0, 1, 254 and 255; line sync is a single pulse at picture element '0' on lines 2–253; picture enable is HIGH on picture elements 20–339 for lines 8–247. The entire video raster contains 256 lines of 360 elements, with the picture occupying 240 lines of 320 elements. If the system displays 50 frames per second the picture element rate is $50 \times 256 \times 360$ which is 4.608 MHz, well within the capability of a PLD.

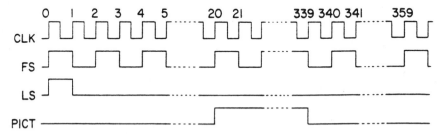

Fig. 9.29 Waveforms for a video controller.

There are many possible ways of building this circuit since the counter requires 17 flip-flops and the decoder 3 outputs, decoding 17 inputs. The minimum chip count solution uses two 22V10s, the first forms a divide-by-360 counter for the line count and also provides the LSB of the frame count. The ten lines are then fed to the other 22V10 which acts as a 7-bit counter to complete the frame, and three outputs are set as combinational to decode the sync pulses. Having already covered counter design at some length, we will just list the equations for the decoder.

$$
\begin{aligned}
\text{LINE} = \ & /\text{L8} * /\text{L7} * /\text{L6} * /\text{L5} * /\text{L4} * /\text{L3} * /\text{L2} * /\text{L1} * /\text{L0} \\
& */(\text{F7} * \text{F6} * \text{F5} * \text{F4} * \text{F3} * \text{F2} * \text{F1} \\
& + /\text{F7} * /\text{F6} * /\text{F5} * /\text{F4} * /\text{F3} * /\text{F2} * /\text{F1}) \\
\text{FRAME} = \ & \text{F7} * \text{F6} * \text{F5} * \text{F4} * \text{F3} * \text{F2} * \text{F1} * /\text{L1} \\
& + /\text{F7} * /\text{F6} * /\text{F5} * /\text{F4} * /\text{F3} * /\text{F2} * /\text{F1} * /\text{L1} \\
/\text{PICT} = \ & /\text{F7} * /\text{F6} * /\text{F5} * /\text{F4} * /\text{F3} \\
& + \text{F7} * \text{F6} * \text{F5} * \text{F4} * \text{F3} \\
& + /\text{L8} * /\text{L7} * /\text{L6} * /\text{L5} * /\text{L4} \\
& + /\text{L8} * /\text{L7} * /\text{L6} * /\text{L5} * \text{L4} * /\text{L3} * /\text{L2} \\
& + \text{L8} * /\text{L7} * \text{L6} * /\text{L5} * \text{L4} * /\text{L3} * \text{L2} \\
& + \text{L8} * /\text{L7} * \text{L6} * /\text{L5} * \text{L4} * \text{L3} \\
& + \text{L8} * /\text{L7} * \text{L6} * \text{L5} * /\text{L4} * /\text{L3}
\end{aligned}
$$

F0–F7 and L0–L8 refer to the frame count and line count bits respectively. A circuit diagram of the whole system is shown in Figure 9.30.

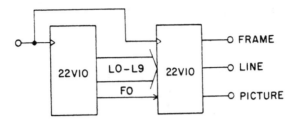

Fig. 9.30 Video controller circuit diagram.

9.3.2.2 Code converters and look-up tables

One of the most common 'logic' uses of PROMs/PLEs is as a look-up table, and a particular example is the code converter. We originally described this application in Section 4.1.5.2 and mention it again here for the sake of completeness. Some functions are available as standard parts from various manufacturers; in particular the BCD-binary conversion, and its converse, are 74S484 and 74S485. Various code conversions such as ASCII–EBCDIC can also be obtained. If the function required is not available as a standard part it will probably be simpler to generate it with a computer program rather than attempt to devise the logic for it.

9.3.2.3 Pseudo-random number generator

Figure 9.31 shows the general schematic of a pseudo-random number generator. It consists of a shift register with feedback to the input generated as a combinational function of two or more outputs. The register should cycle through every possible combination of states, provided that the function is chosen correctly. Depending how complex the function is, a registered PLE, PAL or PLS can be used to build the 'PRN'. We have already described a standard shift register; the logic array can be used to hold the generator function. For example the function:

$$Q8 := Q6 : +: Q5 : +: Q4 : +: Q2$$
$$Q7 := Q8$$
$$Q6 := Q7$$

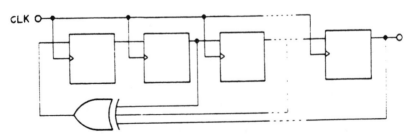

Fig. 9.31 Pseudo-random number generator schematic.

$$Q5 := Q6$$
$$Q4 := Q5$$
$$Q3 := Q4$$
$$Q2 := Q3$$
$$Q1 := Q2$$

gives the following sequence if it starts from FF, once the initial 8 HIGHs have worked through:

A8, 54, AA, D5, EA, F5, 7A, 3D, 9E, CF, 67, ...

The only problem with this function is that it cannot start from 00 or 01 as the feedback will always yield a LOW. As with all state machines, care must be taken to ensure that the initial conditions allow correct operation.

9.3.2.3 *Bidirectional 1-to-8 line mux/demux*

This is an example of a circuit which is a combination of some standard functions described earlier. As such, it could be built from standard TTL or CMOS MSI devices but we will see how these may be replaced by a single PLD. A block diagram is shown in Figure 9.32. If DIR is HIGH then 8-bit wide data (X7–X0) will be multiplexed onto the single data line (Y); if DIR is LOW the single data line will be routed to the parallel port. The address lines, A2–A0, select which of the parallel lines is active, and there is a chip enable, /CE, which makes every output tri-state when HIGH. The equations for this function are:

$$X7.TST = /CE * DIR$$
$$X7 = A2 * A1 * A0 * Y$$

$$X6.TST = /CE * DIR$$
$$X6 = A2 * A1 * /A0 * Y$$

$$X5.TST = /CE * DIR$$
$$X5 = A2 * /A1 * A0 * Y$$

$$X4.TST = /CE * DIR$$
$$X4 = A2 * /A1 * /A0 * Y$$

$$X3.TST = /CE * DIR$$
$$X3 = /A2 * A1 * A0 * Y$$

$$X2.TST = /CE * DIR$$
$$X2 = /A2 * A1 * /A0 * Y$$

$$X1.TST = /CE * DIR$$
$$X1 = /A2 * /A1 * A0 * Y$$

$$X0.TST = /CE * DIR$$
$$X0 = /A2 * /A1 * /A0 * Y$$

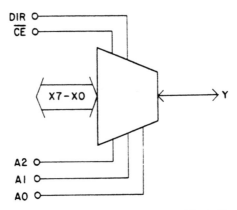

Fig. 9.32 Bidirectional 1-to-8 mux/demux block diagram.

Y.TST = /CE * /DIR
Y = A2 * A1 * A0 * X7
+ A2 * A1 * /A0 * X6
+ A2 * /A1 * A0 * X5
+ A2 * /A1 * /A0 * X4
+ /A2 * A1 * A0 * X3
+ /A2 * A1 * /A0 * X2
+ /A2 * /A1 * A0 * X1
+ /A2 * /A1 * /A0 * X0

This function needs nine bidirectional pins with the ability to support eight AND terms on one of them. This is no problem for a PLA such as the PLS153, but it would require a fairly complex PAL, a 22V10 or EP600 for example, to make this function. A PLE solution is out of the question since they do not possess bidirectional pins.

Several variations of the basic function are possible. There are enough inputs to make two completely independent 4-line to 1-line devices; if a registered device is used, the data can be stored in the output flip-flops. Typical uses for this function are as a simple UART or, for example, to interface to a single-bit device like a dynamic RAM.

9.3.2.5 Crosspoint switch

In a similar vein to the above circuit, the crosspoint switch allows each output to select any input as its data source. The address of the selected input is stored in a latch or register. This design is based on the PLS159 which has enough room for four inputs and four outputs; a larger switch could be accommodated in a larger device, but the equations would be basically the same. An EP900 could support an 8-input 4-output device, for instance, and the MACH110 an 8-input / 8-output switch.

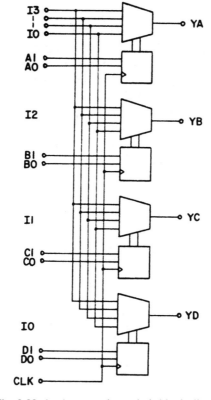

Fig. 9.33 4 × 4 crosspoint switch block diagram.

A block diagram of the device is shown in Figure 9.33 and the equations for the PLS159 are:

$$YA = I3 * QA1 * QA0$$
$$+ I2 * QA1 * /QA0$$
$$+ I1 * /QA1 * QA0$$
$$+ I0 * /QA1 * /QA0$$
$$YB = I3 * QB1 * QB0$$
$$+ I2 * QB1 * /QB0$$
$$\text{etc.}$$
$$LA = LB = 1$$
$$D3 = D2 = D1 = D0 = 1$$

(these are direction terms – not input selection)

The equations for YB, YC, and YD are the same as those for YA with QB, QC, or QD replacing QA. QA, QB, QC and QD are the internally registered A, B, C and D signals.

9.3.2.6 *Majority logic circuit – 1*

In some applications it is necessary to make a decision on the basis of whether there are more HIGH inputs than LOWs, or vice versa. Firstly, we will look at a combinational situation where the inputs are presented in parallel. As an example we can assume that there are five inputs, in which case we want the output to be HIGH if three or more inputs are HIGH. The Karnaugh map of this function is shown in Figure 9.34 and it is apparent that ten AND terms are needed to build this circuit from a standard PAL or PLA. The ten terms are those which include any three of the five inputs; mathematically this is calculated by $^5C_3$. This means that the number of terms will increase rapidly as the number of inputs increases.

	I5	0	0	0	0	1	1	1	1
	I4	0	0	1	1	1	1	0	0
	I3	0	1	1	0	0	1	1	0
I2	I1								
0	0						H		
0	1			H		H	H	H	
1	1		H	H	H	H	H	H	H
1	0			H		H	H	H	

Fig. 9.34 Karnaugh map – 3 out of 5 majority logic circuit.

For more than five inputs, assuming that only odd numbers are considered, it will usually be necessary to use a PLE for this function. The equations for five inputs are:

$$
\begin{aligned}
\text{MAJ} = & \ I5 * I4 * I3 \\
 & + I5 * I4 * I2 \\
 & + I5 * I4 * I1 \\
 & + I5 * I3 * I2 \\
 & + I5 * I3 * I1 \\
 & + I5 * I2 * I1 \\
 & + I4 * I3 * I2 \\
 & + I4 * I3 * I1 \\
 & + I4 * I2 * I1 \\
 & + I3 * I2 * I1
\end{aligned}
$$

9.3.2.7 *Majority logic – 2*

If the data is presented in serial form the problem can be tackled as a sequential circuit. A typical application might be to clean up a noisy signal. If '*n*' samples are taken for each majority decision then the circuit must be

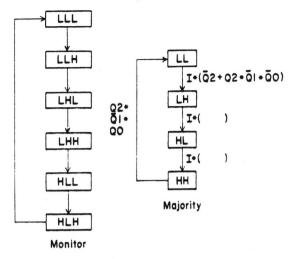

Fig. 9.35 State diagram for sequential 3 out of 5 majority logic circuit.

clocked at a rate $n+1$ times the maximum frequency contained in the signal, or higher. The circuit consists of two counters; one cycles continuously, dividing by six, the other is only incremented when the signal input is high. On the sixth count the output is set HIGH if the second counter is at '3' or more, otherwise it is set LOW. Both counters are reset by the next clock input and the process repeated. The state diagram for this operation is shown in Figure 9.35. This can be put into either a registered PAL or PLS. The more complex solution is the PAL, using D-type flip-flops, so we will quote the equations for this method:

Standard counter:

$$/Q2 := /Q2 * /Q0$$
$$+ Q2 * Q1$$
$$+ /Q1 * Q0$$
$$/Q1 := /Q2 * /Q1 * /Q0$$
$$+ /Q2 * Q1 * Q0$$
$$+ Q2$$
$$/Q0 := = Q0$$
$$+ Q2 * Q1$$

Majority counter:

$$/M1 := /I * /M1$$
$$+ I * /M1 * /M0 * /(Q2 * /Q1 * Q0)$$
$$+ Q2 * /Q1 * Q0$$
$$/M0 := /I * /M0$$
$$+ I * /M1 * M0 * /(Q2 * /Q1 * Q0)$$
$$+ Q2 * /Q1 * Q0$$

Output signal:

$$/O := M1 * M0 * M2 * /Q1 * Q0$$
$$+ O /(* Q2 * /Q1 * Q0)$$

The function '/(Q2 * /Q1 * Q0)' is ideally suited to use of the complement array in a PLS. If a device without a complement array is being used the complement may be replaced by '/Q2 + Q1 + /Q0'.

9.3.2.8 *Stepper motor controller*

This application is described in the AMD PAL Handbook for both half step and full step applications, but the equations are derived for only the full step case. We will derive the half step equations. Figure 9.36 shows the block diagram, with Q1–Q4 driving the motor coils, D defining the direction of rotation, E enabling rotation and S the start-up which loads step-1 into the output register. The step sequence is:

STEP	Q4	Q3	Q2	Q1
1	0	1	0	1
2	0	0	0	1
3	1	0	0	1
4	1	0	0	0
5	1	0	1	0
6	0	0	1	0
7	0	1	1	0
8	0	1	0	0
1	0	1	0	1

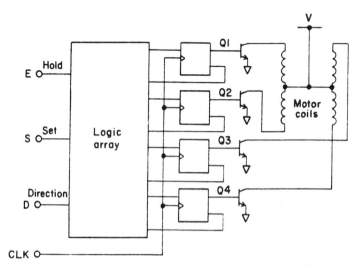

Fig. 9.36 Stepper motor controller block diagram.

Clockwise rotation is obtained from a sequence 1–2–3–4–5–6–7–8–1, anti-clockwise by 1–8–7–6–5–4–3–2–1. Because the PAL example is based on D-type flip-flops we will illustrate this case with J–Ks. Apart from showing an alternative implementation, this has the advantage that the equations can be derived directly from the step table, which is effectively a state table. We can use toggle mode for the waveform generation while setting step-1 uses direct loading; hold mode is implicit in the J–K. The equations are:

$$Q4.J = /D * E * /Q4 * /Q3 * /Q2 * Q1$$
$$+ /D * E * Q4 * /Q3 * Q2 * /Q1$$
$$+ D * E * Q4 * /Q3 * /Q2 * Q1$$
$$+ D * E * /Q4 * /Q3 * Q2 * /Q1$$
$$Q4.K = S * E + \text{the same equations as Q4.J}$$

$$Q3.K = /D * E * /Q4 * Q3 * /Q2 * Q1$$
$$+ /D * E * /Q4 * /Q3 * Q2 * /Q1$$
$$+ D * E * /Q4 * /Q3 * /Q2 * Q1$$
$$+ D * E * /Q4 * Q3 * Q2 * /Q1$$

$$Q3.J = S * E + \text{the same equations as Q3.K}$$

$$Q2.J = /D * E * Q4 * /Q3 * /Q2 * /Q1$$
$$+ /D * E * /Q4 * Q3 * Q2 * /Q1$$
$$+ D * E * /Q4 * Q3 * /Q2 * /Q1$$
$$+ D * E * Q4 * /Q3 * Q2 * /Q1$$
$$Q2.K = S * E + \text{the same equations as Q2.J}$$

$$Q1.K = /D * E * /Q4 * Q3 * /Q2 * /Q1$$
$$+ /D * E * Q4 * /Q3 * /Q2 * Q1$$
$$+ D * E * Q4 * /Q3 * /Q2 * /Q1$$
$$+ D * E * /Q4 * Q3 * /Q2 * Q1$$
$$Q1.J = S * E + \text{the same equations as Q1.K}$$

An alternative solution could have been obtained by entering the state table directly from the step sequence, instead of generating equations. This method, which is more applicable to R–S type PLSs, yields an identical number of AND terms.

9.3.2.9 Shaft encoder

Also associated with motors is the measurement of shaft speed and direction of rotation. The technique often used is a special case of a quadrature detector. Two signals are generated by the rotating shaft, usually by optical means, these are at the same frequency but shifted in phase by 90°. The speed of rotation is determined by the frequency of the signals and the direction by determining which of the two signals is ahead of the other.

Figure 9.37 shows the waveforms for clockwise and anti-clockwise rotation. If the waveforms are sampled between the edges of either of them, a sequence

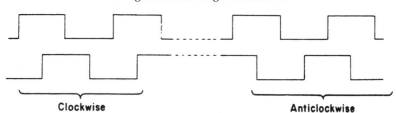

Fig. 9.37 Shaft encoder waveforms.

of HIGHs and LOWs is obtained which can be analysed with synchronous logic to find the speed and direction. The sequences are:

Clockwise		Anti-clockwise	
A	B	A	B
L	L	L	L
H	L	L	H
H	H	H	H
L	H	H	L
L	L	L	L

The waveforms need to be sampled at least four times faster than the maximum signal frequency; the present state of the signals is compared with the previously stored state to obtain the required result. A convenient way of producing the required data is to generate two pulse trains, a clockwise signal which toggles every time an edge is encountered in clockwise rotation, and a similar anti-clockwise signal. Karnaugh maps for driving toggle flip-flops are shown in Figure 9.38, where QA and QB are the stored values of A and B respectively. From these we can derive the following equations:

$$QA := A$$
$$QB := B$$
$$CW.T = /QA * /QB * A * /B$$
$$+ QA * /QB * A * B$$
$$+ QA * QB * /A * B$$
$$+ /QA * QB * /A * /B$$
$$ACW.T = /QA * /QB * /A * B$$
$$+ /QA * QB * A * B$$
$$+ QA * QB * A * /B$$
$$+ QA * /QB * /A * /B$$

9.3.2.10 *Bit rate multiplier*

This is a circuit for producing a waveform whose frequency is proportional to the value of some input number entered in binary code. Operation is based on a Gray code counter; this has the property that only one bit changes state at any one time. If two of the outputs are exclusive-ORed then the resulting

| QB | 0 | 0 | 1 | 1 |
| QA | 0 | 1 | 1 | 0 |
R	A				
0	0				T
0	1	T			
1	1		T		
1	0			T	

CW

| QB | 0 | 0 | 1 | 1 |
| QA | 0 | 1 | 1 | 0 |
B	A				
0	0		T		
0	1			T	
1	1				T
1	0	T			

ACW

Fig. 9.38 Karnaugh maps – shaft encoder.

signal will change state at a rate which is the sum of the two outputs. Referring back to Section 5.2.3.3, we have already seen how to build a Gray code counter. If Q3 toggles at frequency 'f' then Q2 will also toggle at 'f', Q1 at '2f' and Q0 at '4f'. By using the binary inputs to select the appropriate outputs to be exclusive-ORed, any multiple of 'f' from 1 to 7 can be selected. The equation is therefore:

$$O = S2 * Q0 : +: S1 * Q1 : +: S0 * Q2$$

As this is an exclusive-OR function the number of AND terms required doubles with each additional bit, unless extensive use is made of feedback terms. A 256 rate multiplier would need three PLDs without going to complex devices. Figure 9.39 shows an efficient line-up for this function. A PAL20X10 contains a 9-bit counter, a PLE9P8 is a binary to Gray code converter and a PLS173 will make an 8-input exclusive-OR gate, configured as the eight AND gates driving four exclusive-OR gates, which feed the final exclusive-OR.

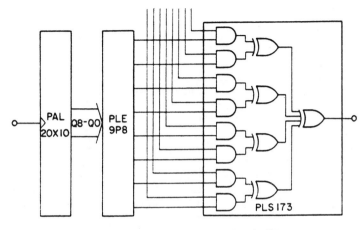

Fig. 9.39 256 bit rate multiplier circuit diagram.

9.3.3   Controllers

9.3.3.1   *Control applications*

Many systems need a controller to modify their operation or to link components of the system together. The essential features of a controller are that they receive signals from the component parts indicating their status, process these signals and then send further signals to inform the components in what way they should modify their operation.

A simple domestic example is a central heating controller. This takes inputs from temperature sensors in the room and hot water tank, and from a clock, and sends signals to the pump and boiler depending on whether the heating is needed or not. While this simple example could be satisfied by combinational logic a state machine solution would allow embellishments to be added without disturbing the basic function. As we shall see, more complex systems need synchronous logic for proper operation.

9.3.3.2   *A general purpose controller*

Figure 9.40 shows what might be considered to be the state diagram of a general purpose controller. The controller rest state is '000__000', on receipt of the start signal 'A' it moves to '001__000' where it waits until 'B' is LOW; thence to '010__000' until 'C' is HIGH. This *soft start* ensures that the process cannot operate until all the components are in the correct state. The initial *run*

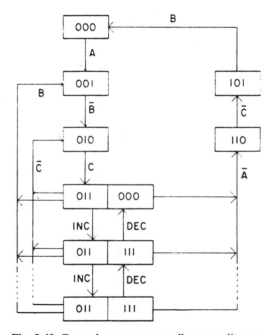

Fig. 9.40 General purpose controller state diagram.

state is '011__000' which can be modified to '011__001', '011__010', etc. according to the various inputs from the system, thereby adapting operation to the requirements of the system. If, during operation, B or C become invalid the system reverts to its soft start state until they return to their proper levels.

The system is turned off by taking 'A' LOW which causes state '110__000' to be entered. When 'C' is LOW it moves to '101__000' until 'B' goes HIGH and the rest state is re-entered. This very generalised picture could be applied to almost any situation where control is required, from gas boilers and arcade games to minicomputers and communication systems via instrumentation and terminals. In this example the first three bits form an underlying level of control to the controller itself while the second three bits define the output from the controller to the system. This is ideally suited to a PLS type of circuit, such as the PLS105 which has a buried register section. This state register can be decoded to give outputs which directly control the system. A typical state table would be:

Inputs					Present state						Next state					
A	B	C	D	E	F5	F4	F3	F2	F1	F0	F5	F4	F3	F2	F1	F0
H	–	–	–	–	L	L	L	L	L	L	L	L	H	L	L	L
H	L	–	–	–	L	L	H	L	L	L	L	H	L	L	L	L
H	L	H	–	–	L	H	L	L	L	L	L	H	H	L	L	L
H	L	H	H	L	L	H	H	L	L	L	L	H	H	L	L	H
H	L	H	H	L	L	H	H	L	L	H	L	H	H	L	H	L
.																
.																
H	L	H	L	H	L	H	H	H	H	H	L	H	H	H	H	L
H	L	H	L	H	L	H	H	H	H	L	L	H	H	H	L	H
.																
H	L	L	–	–	L	H	H	–	–	–	L	H	L	L	L	L
H	H	–	–	–	L	H	–	–	–	–	L	L	H	L	L	L
L	L	H	–	–	L	H	H	–	–	–	H	H	L	L	L	L
L	–	L	–	–	H	H	L	L	L	L	H	L	H	L	L	L
L	H	–	–	–	H	L	H	L	L	L	L	L	L	L	L	L

The inputs 'D' and 'E' are included as typical system inputs which have the effect of changing the second triplet of state bits in a way which is defined according to the needs of the system. In general it is better to define state transitions uniquely rather than use 'don't cares' or logic minimisation. From the practical point of view it makes modification much easier although it may be less efficient in usage of AND terms. From the theoretical point of view it is much safer as there is less risk of opening paths into undefined states or other dead ends. Checking by Karnaugh map will probably be out of the question because of the large number of inputs which would need to be handled.

9.3.3.3 *Handshaking and protocols*

In any system there will be a need to exchange data between parts of the system, or with other systems. In parallel with human conversations, apart from the most casual exchanges, protocols exist to determine who should be talking, when a speaker has finished, and who has precedence to take over. In large 'meetings', a chairman has to make these decisions, while in smaller conversations the protocol is usually understood by each of the participants. While the rules for human conversation tend to be adaptable, the rules governing electronic data exchange are more rigid. A state machine is often the best way to implement the rules.

As an example we can look at a protocol which is familiar to most users of electronic instruments, RS232C. In order for an instrument to converse via an RS232C link the data flow within the instrument must be managed by an internal controller. The architecture which could be used for an RS232C interface is shown in Figure 9.41. It is assumed that the internal processor can only handle data flow in one direction while the interface has separate transmit and receive lines. The controller must ensure that the interface does not transmit and receive simultaneously, and it must manage the *handshake* with the RS232C lines. That is it must send an RTS signal when data is to be sent and look for the CTS response; it must also respond to an incoming RTS signal on its DCD input by sending out DTR when the processor is free to receive data.

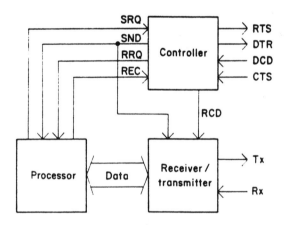

Fig. 9.41 RS232C interface architecture.

Internally the controller is given an SRQ by the processor and sends it SND when all is set up for it to send data. When it receives a request to accept data it sends RRQ to the processor which responds with REC when it is ready; it sends RCD to the transmit/receive circuit at the same time. The state diagram for this is shown in Figure 9.42 and translates into the following state table:

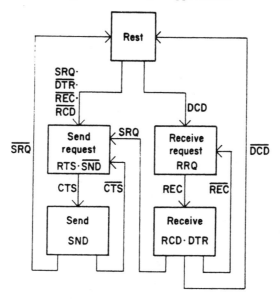

Fig. 9.42 RS232C interface state diagram.

Inputs				Present state					Next state				
SRQ	REC	DCD	CTS	SND	RRQ	RCD	RTS	DTR	SND	RRQ	RCD	RTS	DTR
H	L	L	L	L	L	L	L	L	L	L	L	H	L
H	L	L	H	L	L	L	H	L	H	L	L	H	L
H	L	L	L	H	L	L	H	L	L	L	L	H	L
L	L	L	–	H	L	L	H	L	L	L	L	L	L
–	L	H	L	–	L	L	–	L	–	H	L	–	L
L	H	H	L	–	H	L	–	L	L	H	H	L	H
L	L	H	L	L	L	H	H	L	L	H	L	L	L
–	–	L	L	L	H	–	L	–	L	L	L	L	L

The above table is a simplified version of RS232C sufficient to illustrate the principle of handshaking. In it, the processor is allowed to choose priority between sending and receiving although the controller cannot set DTR HIGH unless CTS is LOW.

9.3.3.4 Multi-processor controller

Many systems now contain several processors sharing resources, such as memory and I/O ports, via a common bus. In order to prevent contention and hogging the bus may be allocated on a time-sharing basis among the processors. This requires a central controller to receive requests for access to the bus and to grant them on a fair basis.

There have been several schemes devised in order to make best use of the resources while allowing a fair distribution of time among the processors. For

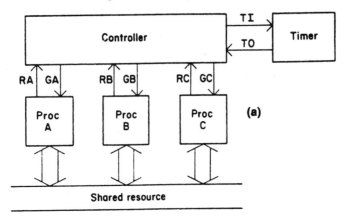

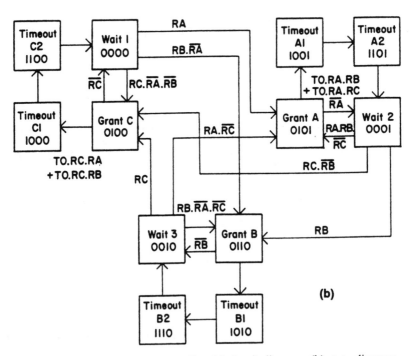

Fig. 9.43 Multi-processor controller: (a) circuit diagram; (b) state diagram.

example, a prioritised system will allow the most important, as judged by the designer, to have first choice even if a lower priority device is in the middle of an operation. This may be satisfactory for some systems, where the roles of the different processors can be evaluated in terms of their importance. A more egalitarian method is the 'round-robin' or last granted lowest priority (LGLP) structure.

LGLP allows each processor a maximum time to use the bus, after which, it checks whether any other processor is requesting access. The last device to use the bus becomes the lowest priority device, but then moves up the pecking order as each processor has a chance to request access. A PLS is again the ideal device to programme for this function. Figure 9.43 shows a PLS in this situation, controlling three processors, and the state diagram for the operation.

The three processors each have a request line to the controller and a grant line back, which allows access when LOW. When a request is granted the timer is started and sends back a timeout signal when time is up, unless the processor relinquishes the bus voluntarily. If another processor is requesting at timeout the grant is removed after three clock pulses and the next grant made. There is always a single clock pulse pause between release and a new grant to avoid contention on the bus. The buried register of the PLS may be used to time the internal delays although the timeout is likely to be too long to use the buried register for that. The state table is therefore:

Inputs				Present state				Next state				Output			
RA	RB	RC	TO	Q3	Q2	Q1	Q0	Q3	Q2	Q1	Q0	GA	GB	GC	TI
H	–	–	–	L	L	L	L	L	H	L	H	L	H	H	L
L	H	–	–	L	L	L	L	L	H	H	L	H	L	H	L
L	L	H	–	L	L	L	L	L	H	L	L	H	H	L	L
–	H	–	–	L	L	L	H	L	H	H	L	H	L	H	L
–	L	H	–	L	L	L	H	L	H	L	L	H	H	L	L
H	L	L	–	L	L	L	H	L	H	L	H	L	H	H	L
–	–	H	–	L	L	H	L	L	H	L	L	H	H	L	L
H	–	L	–	L	L	H	L	L	H	L	H	L	H	H	L
L	H	L	–	L	L	H	L	L	H	H	L	H	L	H	L

(The above terms define round robin priority)

L	–	–	–	L	H	L	H	L	L	L	H	H	H	H	H
–	L	–	–	L	H	H	L	L	L	H	L	H	H	H	H
–	–	L	–	L	H	L	L	L	L	L	L	H	H	H	H

(The above terms define voluntary release)

H	H	–	H	L	H	L	H	H	L	L	H	L	H	H	H
H	–	H	H	L	H	L	H	H	L	L	H	L	H	H	H
H	H	–	H	L	H	H	L	H	L	H	L	H	L	H	H
–	H	H	H	L	H	H	L	H	L	H	L	H	L	H	H
H	–	H	H	L	H	L	L	H	L	L	L	H	H	L	H
–	H	H	H	L	H	L	L	H	L	L	L	H	H	L	H
–	–	–	–	H	L	–	–	H	H	–	–	–	–	–	–
–	–	–	–	H	H	–	–	L	L	–	–	–	–	–	–

(The above terms define timeout)

This table may be extended for more processors or converted to a fixed priority, the latter by using only one group of the priority terms.

9.3.3.5 Dual port RAM controller

A dual port RAM is a random access memory which can be written or read from two places. It allows two systems, or processors, to share a common memory and to communicate using the same storage area. Some dedicated devices are available, but these are limited in size and tend to be rather expensive. If it were possible to use a standard memory then more flexible and cheaper arrangements would be possible. What is needed is a controller circuit which will prevent both sides trying to access the RAM simultaneously and ensure a minimum waiting time if one side tries to access the memory while it is already being used.

A state machine is the ideal way of doing this and a PLS will provide the basis for implementing the function in hardware. The circuit for the dual port RAM is shown in Figure 9.44, along with the state diagram. Buffers are needed for the address and data lines, but these would probably be used anyway so do not represent an additional overhead. Typical buffers would be 74LS244 for the addresses and 74LS245 for the data lines.

Priority has to be built into the design in case of simultaneous requests for access from A and B sides; in this case we will give priority to A. The state diagram allows for a delay between opening the enabled port and granting access, and between removing the grant and closing the port. This makes sure that the lines are settled before access is attempted. The port is left open for a fixed time in each cycle, and at the end of each cycle control is passed to the opposite side if it is requesting access. The state diagram may be converted to a state table as follows:

Inputs				Present state								Next state							
AQ	AW	BQ	BW	AE	AR	AG	BE	BR	BG	D1	D0	AE	AR	AG	BE	BR	BG	D1	D0
H	H	–	–	L	L	L	L	L	L	L	L	H	H	L	L	L	L	L	L
H	L	–	–	L	L	L	L	L	L	L	L	H	L	L	L	L	L	L	L
–	–	–	–	H	–	L	L	L	L	L	L	H	–	H	L	L	L	L	L
L	–	H	H	L	L	L	L	L	L	L	L	L	L	L	H	H	L	L	L
L	–	H	L	L	L	L	L	L	L	L	L	L	L	L	H	L	L	L	L
–	–	–	–	L	L	L	H	–	L	L	L	L	L	L	H	–	H	L	L
–	–	–	–	–	–	H	–	–	–	L	L	–	–	–	–	–	–	L	H
–	–	–	–	–	–	–	–	–	H	L	L	–	–	–	–	–	–	L	H
–	–	–	–	–	–	–	–	–	–	L	H	–	–	–	–	–	–	H	L
–	–	–	–	–	–	–	–	–	–	H	L	–	–	–	–	–	–	H	H
H	H	L	–	H	H	H	L	L	L	H	H	H	H	H	L	L	L	L	L
H	L	L	–	H	L	H	L	L	L	H	H	H	L	H	L	L	L	L	L
L	–	H	H	L	L	L	H	H	H	H	H	L	L	L	H	H	H	L	L
L	–	H	L	L	L	L	H	L	H	H	H	L	L	L	H	L	H	L	L
H	H	–	–	L	L	L	H	–	H	H	H	H	H	L	L	L	L	L	L
H	L	–	–	L	L	L	H	–	H	H	H	H	L	L	L	L	L	L	L
–	–	H	H	H	–	H	L	L	L	H	H	L	L	L	H	H	L	L	L
–	–	H	L	H	–	H	L	L	L	H	H	L	L	L	H	L	L	L	L
L	–	L	–	–	–	–	–	–	–	H	H	L	L	L	L	L	L	L	L

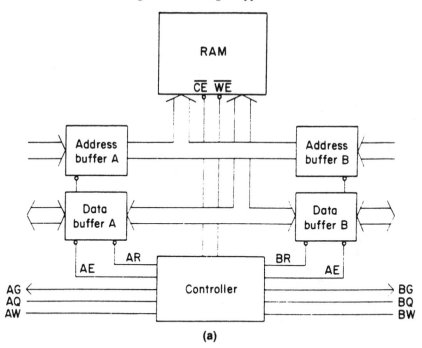

(a)

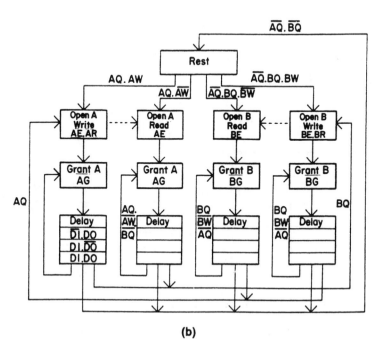

(b)

Fig. 9.44 Dual port RAM controller: (a) circuit diagram; (b) state diagram.

The control signals for the RAM can be derived from the state bits as combinational functions by the following equations:

$$/CE = AE + BE$$
$$/WE = /AR + /BR$$

In the above table the first three lines open the A-port and grant A; the next three lines open the B-port and grant B. The following five lines implement the delay function and then define the routes out of the final delayed state. The last line takes the system back to the rest state if no requests are pending.

9.4 CASE STUDIES

9.4.1 Schematic entry

9.4.1.1 Circuit description

The circuit diagram, shown in Figure 9.45, includes a decoder (74LS138), two quad 2-input NAND gates (74LS00), one and a half hex inverters (74LS04) and a dual D-latch (74LS75) driving various peripheral circuits (RAM, ROM, buffers, etc.). The number of inputs (7) and outputs (8) suggest that this might fit into a standard PLD such as the GAL16V8. In order to do this we must first

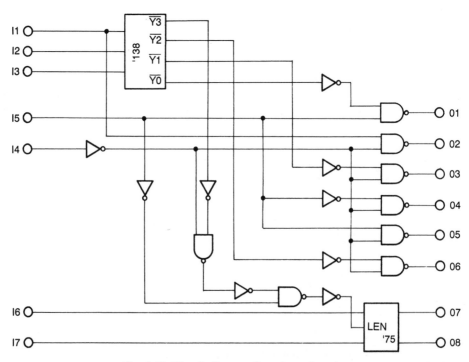

Fig. 9.45 Circuit diagram for schematic entry.

generate the logic equations from the logic diagram. This could be done with a schematic capture programme but, as this is a luxury not available to all designers, we can show how this is done manually.

9.4.1.2   *Derivation of logic equations*

The first step is to redraw the circuit with internal signals labelled and simplified to eliminate the 'double inversions' which often result from using TTL logic, where inverting gates are commoner and faster than non-inverting gates. This is shown in Figure 9.46.

From Section 9.1.5.1 we can write the following equations for the internal signals:

$$Y0 \ = /I1 * /I2 * /I3;$$
$$Y1 \ = /I1 * /I2 * I3;$$
$$Y2 \ = /I1 * I2 * /I3;$$
$$Y3 \ = /I1 * I2 * I3;$$
$$LEN = Y3 * /I4 * /I5;$$

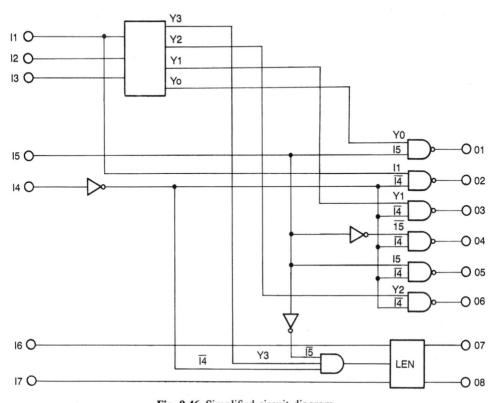

Fig. 9.46 Simplified circuit diagram.

Equations can then be written for the output signals as:

O1 = /(Y0 * I5);
O2 = /(I1 * /I4);
O3 = /(Y1 * /I4);
O4 = /(/I4 * /I5);
O5 = /(/I4 * I5);
O6 = /(Y2 * /I4);
O7 = I6 * LEN + O7 * /LEN + I6 * O7;
O8 = I7 * LEN + O8 * /LEN + I7 * O8;

The equations for O7 and O8 are by reference to Section 9.1.7.1. The equations are written in 'AMAZE' format because a zero-power PLD was needed in this application and the PLC18V8Z was considered to be the most appropriate device. Compilation by 'AMAZE' results in a single product term solution for all outputs except the latches (O7 and O8), which took seven product terms each.

9.4.1.3 Summary

This circuit used nearly all the available I/O resources of the PLD, but less than one third of the logic resources. This is not an unusual result. Nevertheless, five and a half discrete logic chips were condensed into a single PLD, which is quite justifiable economically. This example also shows how a multi-stage logic circuit can be fitted into a two-stage PLD by assigning symbols to the internal logic functions and allowing a logic compiler to expand these as part of the normal compilation process.

9.4.2 Multi-function sockets

9.4.2.1 System description

One major advantage of PLDs is that their function can be changed, and with it the function of the system they are in, without having to make changes to the PCB layout. One simple example is in a decoding circuit where larger memories can be addressed merely by changing the programming of the address decoder. This property was exploited to an even greater extent in the following example.

The system being designed was an analogue signal switching complex with a remote front panel. The front panel switches controlled relays which routed the analogue signals between various input and output ports. It was planned that the same PCBs could be used for all the different switching arrangements and their associated front panels, with a small cable linking the two locations. Figure 9.47 shows the basic arrangement.

9.4.2.2 System implementation

It was clear from the start that a state machine would be needed to control the front panel. As an example of the type of function to be implemented, one

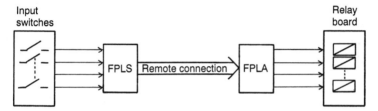

Fig. 9.47 Block diagram of relay control circuit.

variation called for the monitoring of six audio channels. Each channel could be selected by a momentary action switch which also caused all output to be muted until released; this prevented switching 'clicks' from being heard. A simple on-off switch then enabled two more momentary action switches which selected between normal stereo, phase reversed or the mono effect of either. Switching between channels could only occur when mode selection was disabled. The basic state diagram for this arrangement is shown in Figure 9.48.

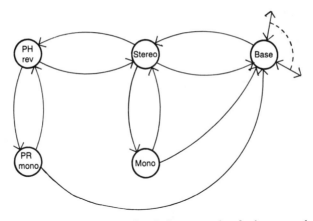

Fig. 9.48 State diagram for design example of relay control.

The momentary action switches cause an added complication for they cannot be used for direct state switching. If a HIGH is the condition for a jump from A to B and, on subsequent release and remake, from B to A, an intermediate state must also be specified for each transition; the final state can only be entered on release of the switch. The right and wrong way of doing this is shown in Figure 9.49. The penalty for using the wrong method is for the state machine to oscillate between A and B, the final state being determined only by the moment at which the switch is released. The advantage, in this application, is that the intermediate state also defines the mute operation.

The provision of intermediate states means that 72 states have to be defined in the state machine. They also have to be decoded in order to operate the

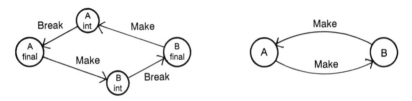

Fig. 9.49 Momentary switching state diagrams.

correct relays, or combination of relays. Decoding is, as we have seen, one of the primary functions of combinational PLDs. The state machine outputs were, therefore, used as the signals for communication between the front panel and the remote relay cards. Decoding was performed by combinational PLDs on the relay cards.

Note that, although one specific function for the switching has been described, any function involving up to 12 switches of any type can be controlled by a logic sequencer on the front panel. With eight bits in the state register, up to 256 relays can be controlled on the remote card. The only limitations are the complexity of the logic which can be accommodated in the sequencer and the number of decoders which can be placed on the relay card.

9.4.2.3 *Defining the state machine*

By judicious grouping of function bits, the transitions between all 72 states can be defined in just 18 transition terms. The six channels take three bits to define them and, in the actual system, were just encoded according to their binary value. Discounting the intermediate states, each channel could be in one of five states, thus another three bits were needed to define this. All intermediate states were defined by a HIGH on a seventh bit, final states by a LOW.

The 'base' states for each channel were simply called [BASE1], [BASE2] etc. and a state [BASE] was defined independent of channel number. The stereo [ST], phase reverse [PR], mono [MC] and phase reverse mono [PM] states were also defined without specifying channel. In each case the intermediate state was assumed. Two state definitions [INT] and [FIN] were then used irrespective of both channel number and channel state.

The switch inputs were also given input vector definitions. [S1] thus defined the switch for channel 1 HIGH and all others LOW; the other channel switches were defined as [S2], [S3] and so on. The function enable was called [S7], switching into and out of phase reverse was [S8], but had to include the function enable HIGH as well; similarly [S9] vector had both mono switch and function enable HIGH. A final switch definition [S0] was defined as all switches, except function enable, LOW. [S00] was just function enable LOW.

The state transitions were then:

WHILE [BASE]
   IF [S1] THEN [BASE1]
   IF [S2] THEN [BASE2]
   IF [S3] THEN [BASE3]

```
      IF [S4] THEN [BASE4]
      IF [S5] THEN [BASE5]
      IF [S6] THEN [BASE6]
      IF [S7] THEN [ST]

WHILE [ST]
      IF [S8] THEN [PR]
      IF [S9] THEN [MC]
      IF [S00] THEN [BASE]

WHILE [PR]
      IF [S8] THEN [ST]
      IF [S9] THEN [PM]
      IF [S00] THEN [BASE]

WHILE [MC]
      IF [S9] THEN [ST]
      IF [S00] THEN [BASE]

WHILE [PM]
      IF [S9] THEN [PR]
      IF [S00] THEN [BASE]

WHILE [INT]
      IF [S0] THEN [FIN]
```

By making each state a combination of three 'mini-states', in this case channel number, switch function and intermediate/final, the transition conditions can be simplified so that 162 transitions are defined in 18 equations.

9.4.2.4 *Summary*

Two conclusions may be derived from this example. First the point that PLDs can be used to define the function of a general purpose circuit board. This ranks them as equivalent to microprocessors which need a stored program to make them suitable for a specific application. Indeed a microprocessor would have been an alternative solution in this application, but it would have been bulkier, more time consuming to reprogram and, probably, more expensive.

The second lesson is with respect to defining states. Some forethought about how states are to be numbered can save much time and trouble in defining transitions. This is fairly obvious in the example above, but it should be carefully thought out in any state machine design.

9.4.3 **Multi-chip state machine**

9.4.3.1 *System description*

The final case study is an example of a system where the synchronous logic could not be fitted into a single sequencer, although it might now be a candidate for an LSI PLD.

The unit is a display panel which accepts warning signals and alerts an operator with a flashing light and buzzer. Each warning has an associated lamp and response switch – momentary operation again. Activation of the switch changes the lamp to continuous and the buzzer to intermittent. These conditions remain until the warning disappears (i.e. corrective action taken) or a second warning arrives, when the first lamp remains continuous, the second lamp flashes and the buzzer returns to continuous operation. The system was designed to accommodate a large but unspecified number of warning signal lines.

9.4.3.2 *State diagram derivation*

Two facts were soon apparent; each lamp and the buzzer would have an individual state diagram, but these would have to interact via global signals derived from the lamp states, and which could be used to 'carry' information between PLDs. The state diagram for one lamp is a simple circular diagram with just four states, as in Figure 9.50. If the warning signal and response button are called WS and RB respectively, we can write state equations as follows:

WHILE [INACT]
  IF WS THEN [ACTIVE]

WHILE [ACTIVE]
  IF RB THEN [FLASHI]

WHILE [FLASHI]
  IF /RB THEN [FLASHF]

WHILE [FLASHF]
  IF /WS THEN [INACT]

The four states can be defined by two bits (say L1 and L0) and, again, by judicious choosing the remaining logic can be simplified. States [FLASHI] and [FLASHF] correspond to a flashing lamp condition and may both contain L1.

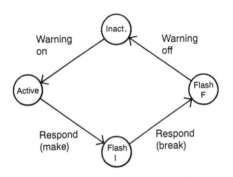

Fig. 9.50 Warning lamp state diagram.

The equation for the lamp output is therefore:

$$LO = /L1 * L0 + L1 * INT$$

where INT is an intermittent HIGH signal defining the flash period. The state diagram for the buzzer can now be drawn, as in Figure 9.51. LA is a signal

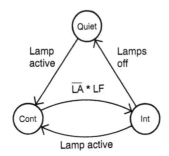

Fig. 9.51 Buzzer state diagram.

which is HIGH when any lamp is on, and is just all the /L1 * L0 terms OR-ed together; LF is a signal which is HIGH when any lamp is flashing, and is formed by OR-ing all the L1 functions. Thus the transition equations are:

WHILE [QUIET]
 IF LA THEN [CONT]

WHILE [CONT]
 IF /LA * LF THEN [INTER]

WHILE [INTER]
 IF LA THEN [CONT]
 IF /LF THEN [QUIET]

The buzzer equation can be derived from the state bits in a similar manner to the lamps.

9.4.3.3 Hardware implementation

Only one chip needs to contain a buzzer control state machine, the rest need only have warning signal channels as their status can be passed on by the LA and LF signals. Each channel needs two inputs, two flip-flops which can be internal, and one output; in addition the chip needs inputs for LA, LF and INT signal plus a clock, and outputs for passing on LA and LF. Five channels would thus fit into a 24-pin PLD such as the PLC42VA12 or PAL32VX10, which allow their flip-flops to be buried.

Larger PLDs would naturally accommodate more channels, and inspection shows that the number is limited by I/Os rather than logic cells. For example, an EPM5128 could contain 18 channels but that would use only 36 of its 128 flip-flops. A block diagram of the hardware solution is shown in Figure 9.52.

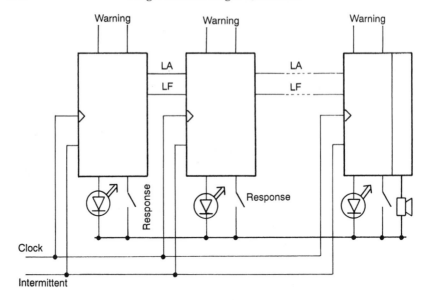

Fig. 9.52 Block diagram of hardware solution.

9.5 SUMMARY

The applications listed in this chapter are a small selection of what is possible with PLDs. We hope that the earlier sections will assist in putting your ideas into practice, and that the later sections may generate fresh ideas and also help with solving some of the problems which may be encountered with more complex designs.

The reader is also advised to consult the handbooks published by the device manufacturers. These tend not to be unbiased towards particular products but also contain many useful hints and ideas.

Conclusion

Although technology has moved on, there is very little that needs changing from the conclusion to the first edition. Memories are available with eight megabits, gate arrays can be obtained with over 100 000 gates and microprocessors with 32 bits are commonplace. Programmable logic has also increased in complexity, but has further to go.

Prophesy is always dangerous, but many of the 'predictions' of the first edition have come to pass, or are still on course. New companies have appeared with innovative architectures and technologies, but some of the old ones have disappeared, either from programmable logic or completely. The most remarkable loss was Monolithic Memories who introduced PALs; MMI were absorbed into Advanced Micro Devices who might now be considered as market leaders.

Certainly CMOS has emerged as the predominant technology for new devices. A few bipolar devices are still being promoted at the high-speed end of the market but no new bipolar architectures have emerged in the last five years. The new architectures are moving away from traditional PAL/PLA structure, as I suggested, and there is every reason to suppose that this trend will continue. The one area which has not been entered to a large degree is that of low power. Some simple architectures are available in 'zero-power' versions but makers of portable equipment are still very limited in their choice of programmable device. An order of magnitude reduction in power consumption without an order of magnitude increase in price would be very welcome in this area.

Software support has as much of its innovative effort involved in keeping up with new architectures as with new design features. However, the universal packages are very transparent to the designer, even to the extent of recommending suitable target devices and allocating device pins to named signals. It is difficult to see what improvements could be made in this area except in the FPGA field. Here we are still reliant on dedicated place and route software for completing designs. If a manufacturer could bring out a universal place and route to interface to his design package, at a reasonable cost, it might cause a step function increase in the use of these devices.

As far as competitive technologies are concerned, the cost 'penalty' against discrete logic is gradually eroding. A standard PLD is now only four or five times the cost of a TTL gate, and comparable with the price of one of the more complex MSI functions. My original prediction of 1996 for parity may not be far wrong, although power consumption may still prove to be a barrier for complete replacement.

The start-up costs, or NRE, of masked devices still operates in favour of programmable devices, although e-beam technology and increased automation are helping to lower the cost and minimum batch size for a masked ASIC. Some companies are promoting PLDs as a development tool for masked ASICs, using them to prove both the design and market before changing to an identical masked device. This is certainly a feasible approach and will help to cut costs of full production units. In spite of the comments in the first edition, it is clear that masked devices will always be significantly cheaper than their programmable equivalent.

The trend towards integrating PLDs with other functions is one which is likely to continue. This approach, together with the obvious increase in size and speed which is likely, will, I am sure, be the way in which PLDs develop. As always it is you, the user, who dictates new developments and I hope that this book has contributed in some way to that end.

Appendix 1
PLD Manufacturer Details

A broad summary of the ranges supplied by various manufacturers is given in Table 12.

MANUFACTURER	PLEs	Std. PALs	Comp. PALs	FPLAs	FPLSs	LSI PLDs	FPGAs	ECL PALs
Actel							X	
Altera			X			X	X	
Atmel	X		X			X		
A.M.D.	X	X	X		X	X	X	X
Cypress	X		X		X	X		
Exel	X				X			
I.C.T.	X		X	X			X	
Intel	X		X			X		
Lattice			X		X			
National Semiconductor	X	X	X		X	X		X
Philips/Signetics	X	X	X	X	X	X		X
Plessey							X	
Plus Logic						X		
S.G.S./Thomson	X		X					
Texas Instruments	X	X	X		X	X	X	X
Xilinx							X	

Table 12

Details of each manufacturer including addresses in USA and Europe and relevant literature and software support are listed below.

ACTEL CORPORATION

USA – 955 E. Arques Avenue, Sunnyvale, CA 94086; phone: (408) 739-1010

UK (for all Europe) – Intec 2, Studio 3, Wade Road, Basingstoke, Hants RG24 0NE; PHONE: 0256 29209

Software support – Action logic system

Programming support – Activator programming system

Literature – Databook and design guide

ALTERA CORPORATION

USA – 2610 Orchard Parkway, San Jose, CA 95134-2020; phone: (408) 984-2800; telex: 888496; fax: (408) 248–7097

UK – 21 Broadway, Maidenhead, Berkshire SL6 1JK; phone: 0628 32516; telex: 94016389; fax: 0628 770892

France – 72-78 Grande Rue, 92310 Sevres; phone: 1.45.34.3787; fax: 1.45.34.0109

Germany – Ismaninger Strasse 21, 8000 Munchen 80; phone: 89/413.00.6-14; fax: 89/470.62.84

Belgium (European HQ) – 25 Avenue Beaulieu, 1160 Bruxelles; phone: 2-660.20.77; telex: 27087901; fax: 2-660.52.25

Japan – Ichikawa Gakugeidai Building, 2nd floor, 12-8 Takaban 3-chome, Meguro-ku, Tokyo 152; phone: 03-716-2241; fax: 03-716-7924

Software support – A + PLUS (EPLDs), MAX + PLUS (MAX)

Programming support – PLE3-12A programming unit

Literature – Databook (includes applications)

ATMEL CORPORATION

USA – 2125 O'Nel Drive, San Jose, CA95131; phone: (408) 441-0311; fax: (408) 436-4200

UK (European HQ) – Coliseum Business Centre, Riverside Way, Camberley, Surrey GU15 3AQ; phone: 0276 686677; fax: 0276 686697

Software support – Atmel-ABEL

Literature – Databook (includes applications)

ADVANCED MICRO DEVICES INC. (AMD)

USA – 901 Thompson Place, P.O. Box 3453, Sunnyvale, CA94088-3453; phone: (408) 732-2400; telex: 34-6306

UK – AMD House, Goldsworth Road, Woking, Surrey GU21 1JT; phone: 0483 740440; telex: 859103; fax: 0483 756196

France – (address not known); phone: 01-49-75-10-10; telex: 263282; fax: 01-49-75-10-13

Germany – Rosenheimer Strasse 143B, 8000 Munchen 80; phone: 089 41140; telex: 523883; fax: 089 406490

Japan – (address not known); phone: 03 345-8241; telex: 24064; fax: 03 342-5196

Software support – PALASM4 (PALs and PLSs), PGA software (FPGAs)

Programming support – LabPro (PALs and PLSs), AmPGA081 (serial PROMs for FPGAs)

Literature – Databooks for PLDs and MACH family, and MACH family casebook

CYPRESS SEMICONDUCTOR

USA – 3901 N. First Street, San Jose, CA95134; phone: (408) 943-2600; telex: 821032; fax: (408) 943-2741

UK – 3 Blackhorse Lane, Hitchin, Herts SG4 9EE; phone: 0462 420566; fax: 0462 421969

France – Miniparc Bat. no 8, Avenue des Andes 6, Z.A. de Courtaboeuf, 91952 Les Ulis Cedex; phone: 01-69-07-55-46; fax: 01-69-07-55-71

Belgium (European HQ) – Avenue Ernest Solvay 7, B-1310 La Hulpe; phone: 02-652-0270; telex: 64677; fax: 02-652-1504

Japan – Fuchu-Minami Bldg. 2F, 10-3 1-Chome, Fuchu-machi, Fuchu-shi, Tokyo; phone: 423-69-82-11; fax: 423-60-82-10

Literature – Databook and handbook

EXEL MICROELECTRONICS

USA – 2150 Commerce Drive, P.O. Box 49038, San Jose, CA95161; phone: (408) 432-0050; fax: (408) 432-8710

UK – Rohm Electronics UK, 15 Peverel Drive, Granby, Milton Keynes, MK1 1NN; phone: 0908 271311; fax: 0908 270380

France – Rohm Electronics GMBH, Bureau de Liaison, 24 Rue Saarinen Silic 224, 94528 Rungis Cedex; phone: 01-46-75-90-51; fax: 01-46-75-00-47

Germany – Rohm Electronics GMBH, Muhlenstrasse 70, D-4052 Korschen-brioch 1; phone: 2161-6101-35; fax: 2161-6421-02

Japan – K H Electronics Corp., Landic No. 2, Akasaka Bldg. 3 Flr, 10-9 Akasaka, 2-Chome Minato-ku, Tokyo; phone: 03-587-1041; fax: 03-584-6394

Software support – AdET

Programming support – E2 programming system

Literature – Databook (includes applications)

INTERNATIONAL CMOS TECHNOLOGY INC.

USA – 2125 Lundy Avenue, San Jose, CA95131; phone: (408) 434-0678; fax: (408) 434-0688

UK – Sequoia Technology Ltd., Unit 5 Bennet Place, Bennet Road, Reading, Berks RG2 0QX; phone: 0734 311822; fax: 0734 312676

France – (addresses not known)
ASAP phone: 01-30-43-82-33
MISIL phone: 01-45-60-00-21

Germany – (address not known)
United Semiconductor Engineering phone: 89-33-92-92
Alfred Neye Enatechnik phone: 41-06-61-20

Far East – Excel Associates Ltd., 1502 Austin Tower, 22-26A Austin Avenue, Tsimshatsui, Kowloon, Hong Kong; phone: 03-7210900; fax: 03-696826

Software support – APEEL (PEEL devices), PLACE (FPGAs)

Programming support – PDS1 programmer

Literature – Databook PEEL software and applications, PLACE software and applications

INTEL CORPORATION

3065 Bowers Avenue, Santa Clara, CA95051; phone: (800) 548-4725 (literature), (800) 538-1876 (enquiries)

UK – Pipers Way, Swindon, Wilts SN3 1RJ; phone: 0793 696000; fax: 0793 641440

France – 1 Rue Edison – BP 303, 78054 St Quentin en Yvelines Cedex: phone: 01-30-57-70-00; fax: 01-30-64-60-32

Germany – Dornacher Strasse 1, 8016 Feldkirchen bei Munchen; phone: 089 90992-0; fax: 089 9043948

Japan – 9 offices e.g. Daiichi Mitsugi Bldg., 1-8889 Fuchu-cho, Fuchu-shi, Tokyo 183; phone: 0423-60-7871; fax: 0423-60-0315

Software support – iPLS II

Programming support – Intel APT

Literature – Handbook (includes data and applications information)

LATTICE SEMICONDUCTOR CORPORATION

USA – 5555 N.E. Moore Ct., Hillsboro, Oregon 97124; phone: (503) 681-0118; telex: 277338; fax: (503) 681-3037

In the rest of the world, Lattice are represented by several distributors in each country:

UK – Macro-Marketing phone: 0628 604383
Micro Call phone: 084 421 5405
Silicon Concepts phone: 0428 77617

France – Aquitech phone: 01-40-96-94-94
Franelec phone: 01-69-20-20-02
Tekelec Airtronics phone: 01-45-34-75-35

Germany – Alfatron Gmbh phone: 89 329 0990
Bacher Gmbh phone: 89 773081

Japan – Ado Electronic Indust. Co. phone: 03 257 2630
Chemi-Con Sales Co. phone: 03 788 7541
Japan Macnics Corp. phone: 044 711 0022
Hoei Denki phone: 03 293 2401

Literature – Databook

NATIONAL SEMICONDUCTOR CORPORATION

USA – 2900 Semiconductor Drive, P.O. Box 58090, Santa Clara, CA95052-8090; phone: (408) 721-5000

UK – The Maple, Kembrey Park, Swindon, Wilts SN2 6UT; phone: 0793 614141; telex: 444674; fax: 0793 697522

France – Centre d'Affaires La Boursidiere, Batiment Champagne B.P.90, Route National 186, F-92357 Le Plessis Robinson; phone: 01-40-94-88-88; telex: 631065 fax. 01-40-94-88-11

Germany – Industriestrasse 10, D-8080 Furstenfeldbruck; phone: 081-41-103-0; telex: 527649; fax: 081-41-103554

Japan – Sanseido Bldg. 5F, 4-15 Nishi Shinjuku, Shinjuku-ku, Tokyo 160; phone: 03-299-7001; fax: 03-299-7000

Software support – OPAL

Literature – Databook and design guide, PLD toolkit

PHILIPS SEMICONDUCTORS

Netherlands – Postbus 90050, 5060 PB Eindhoven; phone: 040 783749

UK – Mullard House, Torrington Place, London WC1E 7HD; phone: 071-580 6633; fax: 071-436 2196

USA – Signetics, 811 East Arques Avenue, Sunnyvale, CA94088-3409; phone: (408) 991-2000

France – 117 Quai du President Roosevelt, 92134 Issy-les-Moulineaux Cedex; phone: 01-40-93-80-00; fax: 01-40-93-86-92

Germany – Burchardstrasse 19, D-2 Hamburg; phone: 040-3296-0; fax: 040-3296-912

Japan – Philips Bldg. 13-37, Kohnan 2-chome, Minato-ku, Tokyo 108; phone: 03-813-3740-5028; fax: 03-813-3740-0570

Software support – AMAZE (PALs and FPLAs), SNAP (FPLSs and PML)

Literature – Data handbook (includes application notes)

PLESSEY SEMICONDUCTORS LTD.

UK – Cheney Manor, Swindon, Wilts SN2 2QW; phone: 0793 518000; telex: 449637; fax: 0793 518411

USA – Sequoia Research Park, 1500 Green Hills Road, Scotts Valley, CA95066; phone: (408) 438-2900; telex: 4940840; fax: (408) 438-5576

France – (address not known) phone: 64-46-23-45; telex: 602858; fax: 64-46-06-07

Germany – (address not known) phone: 089 3609 06-0; telex: 523980; fax: 089 3609 06-55

Singapore (S.E. Asia HQ) – (address not known) phone: 2919291; fax: 2916455

Software support – ERA development system (includes emulator)

Literature – Data sheets

PLUS LOGIC INC.

USA – 1255 Parkmoor Avenue, San Jose, CA95126; phone: (408) 293-7587; fax: (408) 293-7587

Germany (European HQ) – Aidenbachstrasse 137B, D-8000 Munchen 71; phone: 089-78-45-44; fax: 089-78-09-926;

UK distributor – Abacus Electronics, Bone Lane, Newbury, Berks RG14 5SF; phone: 0635 36222; fax: 0635 38670

Software support – PLUSTRAN

Literature – Data sheets and application notes

SGS-THOMSON MICROELECTRONICS

SGS-Thomson have several sales offices in many European countries: just one example will be given per country.

Italy – Via le Milanofiori, Strada 4, Palazzo A/4/A, 20090 Assago (MI); phone: 39-2-89213-1; telex: 330131; fax: 39-2-825449

UK – Planar House, Parkway, Globe Park, Marlow, Bucks SL7 1YL; phone: 0628 890800; telex: 847458; fax: 0628 890391

USA – 1000 East Bell Road, Phoenix, AZ85022-2699; phone: (602) 867-6100

France – 7 Avenue Gallieni BP93, 94253 Gentilly Cedex; phone: 01-47-40-75-75; telex: 632570; fax: 01-47-40-79-10

Germany – Gutleutstrasse 322, 6000 Frankfurt; phone: 69-237492; telex: 176997 689; fax: 60-231957

Japan – Nisseki-Takanawa Bld. 4F, 2-18-10 Takanawa, Minato-ku, Tokyo 108

Literature – Databook (includes applications)

TEXAS INSTRUMENTS INC.

USA – PO Box 225012, Dallas, Texas 75265; phone: (214) 995-6531

UK – Manton Lane, Bedford, MK41 7PA; phone: 0234 270111; telex: 82178

France – 8-10 Avenue Morane Saulnier BP67, 78141 Velizy Villacoublay Cedex; phone: 01-30-70-10-03; telex: 698707

Germany – Haggertystrasse 1, 8050 Freising; phone: 081 61/80-0; telex: 526529

Japan – several offices, for example:
Aoyama Fuji Bldg., 3-6-12 Kita-Aoyama Minato-ku, Tokyo 107; phone: 03-3498-2111

Software support – EP APLUS (EPLDs), Action logic system (FPGAs – includes programming)

Literature – Databook and datasheets (includes applications)

XILINX INC.

USA – 2069 Hamilton Avenue, San Jose, CA95125; phone: (408) 559-7778; fax: (408) 559-7114

UK (European HQ) – Station House, Bepton Road, Midhurst, Sussex GU29 2RE; phone: 0730 816725; fax: 0730 814910

Japan – Okura and Co., 6-12 Ginza Nichome, Chuo-ku, Tokyo 104; phone: 03-566-6361; fax: 03-563-5447

Software support – XACT development system

Programming support – XC-DS81 programmer

Appendix 2
Software and Programmer Manufacturers

Most of the device manufacturers listed in Appendix 1 provide software and programming support for their PLDs. Many also recommend 'third-party' products to support their devices. In Appendix 2 we have listed the major suppliers of software and programmers with, where available, a brief description of the products offered.

ACCEL TECHNOLOGIES INC.

USA – 6825 Flanders Drive, San Diego, CA92121; phone: (619) 554-1000; fax: (619) 554-1019;

UK – Computer Solutions Ltd., Canada Road, Byfleet, Surrey KT14 7HQ

TangoPLD is a PLD compiler incorporating the ESPRESSO minimizer and test vector generation via simulation. Also available is Tango-Schematic for schematic logic entry. Target devices include PALs, PLAs, GALs, PEELs and EPLDs. Will install on any PC with 640k RAM and a hard disk.

B P MICROSYSTEMS

USA – 10681 Haddington, Suite 190, Houston, TX77043; phone: (713) 461-9430; fax: (713) 461-7413

The CP-1128 is driven from the parallel printer port of a PC. The 28-pin socket allows any standard PLD up to that size to be programmed. Vector testing and fuse map editing is also supported.

CAPLIANO COMPUTING

Canada – PO Box 86971, North Vancouver, BC V7L 4P6; phone: (604) 669-6343

Capliano provide support for many design tools, such as ABEL (see Data I/O), on Apple Macintosh systems.

CITADEL PRODUCTS LTD

UK – 50 High Street, Edgware, Middlesex HA8 7EP; phone: 081-951 1848

The PC-82 plugs into an expansion slot of a PC and covers a basic range of PALs, GALs, FPLAs, PEELs and EPLDs. With adaptors, some gang programming and LSI PLD programming is possible. Device testing is also covered by the PC-82.

DATA I/O CORPORATION

USA – 10525 Willows Road N.E., PO Box 97046, Redmond, WA98073-9746; phone: (206) 881-6444; fax: (206) 882-1043

UK – 660 Eskdale Road, Winnersh, Wokingham, Berks RG11 5TS; phone: 0734 440011; fax: 0734 448700

Germany – Lochhamer Schlag 5A, 8032 Graefelfing; phone: 089 858580

Netherlands (European HQ) – World Trade Centre, Strawinskylaan 537, 1077 XX Amsterdam; phone: 020 6622866

Japan – Sumitomoseimei Higashishinbashi, Building 8F, 2-1-7 Higashi-Shinbashi, Minato-ku, Tokyo 105; phone: 011-813-3432-6991

Data I/O provide an extensive range of software and programming products; to summarise:

ABEL-4 PLD is a design software which covers, in principle, every PLD architecture except large FPGAs. It runs on a PC with 640k RAM although some of the advanced features may require up to 8kbytes of RAM. Design entry can be in equation, truth table or state machine format, and it will accept output from standard schematic entry packages. It features design optimisation and the SmartPart database allows automatic device selection after logic compilation. Device simulation includes fault grading and testability analysis before test vector generation.

ABEL-FPGA includes all the features of ABEL-4 but adds full FPGA capability. FPGA designs are output in suitable formats to interface directly to manufacturers' place and route software.

FutureNet is a schematic entry package offered by Data I/O for direct interface to ABEL.

UniSite is a universal programmer with up to 84-pin drivers, making it suitable for all PLDs up to the largest FPGAs. Package variations are catered for by a unique 'matchbook' system which does away with the need for separate programming adaptors.

Other programmers include the 2900 system, which covers virtually all PLDs up to 40 pins, and the 212 programmer which is targeted at CMOS PLDs in 20 and 24-pin packages.

DATAMAN

UK – Station Road, Maiden Newton, Dorset DT2 0AE; phone: 0300 20719; telex: 418442; fax: 0300 21012

Omni-Pro II is a low cost universal programmer driven from PC expansion slot. With a 40 pin socket, it supports 20 and 24 pin bipolar and CMOS PLDs and, with adaptors, a small range of LSI devices.

DIGELEC INC.

Europe – Brudermuhlstrasse, 8000 Munich 70, Germany; phone: 089 776098

USA – 20144 Plummer Street, Chatsworth, CA91311; phone: (800) 367-8750

System UP-803 is a stand-alone universal programmer with a limited built-in logic compiling and editing ability. The newer 860 system is a portable stand-alone universal programmer with remote control capability. It supports PLDs up to 40 pins, including many LSI devices, and is upgradable via interchangeable modules.

ELAN DIGITAL SYSTEMS LTD.

UK – Elan House, Little Park Farm Road, Segensworth West, Fareham, Hants PO15 5SJ; phone: 0489 579799; fax: 0489 577516

USA – 538 Valley Way, Milpitas, CA95035; phone: (408) 946-3864

EF-PER series programmers use a series of modules (ZIFPACs) which plug into a master control unit, the model 5000. At present only CMOS PLDs are covered by this system. The 1000 series programmer supports bipolar devices in one of its modules.

HI-LO SYSTEM RESEARCH CO.

Taiwan – Room 604, 6F No. 2 Lane 995, Ming Shen E. Road, Taipei; phone: 02-7640215; telex: 24071; fax: 02-7566403

UK – Nohau UK Ltd., The Station Mill, Alresford, Hants SO24 9JG; phone: 0962 733140; fax: 0962 735408

ALL-03 is a universal programmer controlled by a PC via an expansion slot. the 40-pin socket is driven by pin drivers which can configure almost any programming algorithm. All the standard 20 and 24-pin bipolar and CMOS devices are covered and many 40-pin PLDs, although these may need additional adaptors.

ICE TECHNOLOGY LTD.

UK – Unit 4, Penistone Court, Station Building, Penistone, South Yorkshire S30 6HG; phone: 0226 767404; fax: 0226 370434

The Speedmaster 1000 is a universal programmer which is driven from the parallel port of a PC. A 40-pin socket means that a large range of PLDs up to 40 pins is supported, including some LSI parts although these need separate adaptors.

INLAB INC.

USA – 2150-1 West 6th Avenue, Broomfield, CO80020; phone: (303) 460-0103

No details are to hand of the programmers made by this company.

ISDATA GMBH

Germany – Daimlerstrasse 51, W-7500 Karlsruhe 21; phone: 07 21 75 1087; fax: 07 21 75 2634

USA – 800 Airport Road, Monterey, CA93940; phone: (408) 373-7359

LOG/iC is a powerful compiler for PLDs and FPGAs. Logic information may be presented as Boolean equations, VHDL files, state diagrams, or a mixture of these as well as interfaces to many standard schematic capture packages. Optimisation can be performed by proprietary methods or by the industry standard ESPRESSO algorithm. Partitioning is interactive, giving potentially denser solutions as the designer can allocate resources within targets in the most efficient manner. Device selection is software aided so the designer can select from the widest range of targets. Finally there is an interactive waveform simulator for design verification and test vector generation.

JAPAN MACNICS CORPORATION

Japan – 516 Imaiminami-Cho, Nakahara-ku, Kawasaki City 211; phone: 044-711-0022

The Promac 11 is a stand alone universal programmer with PLD support up to 28-pin complexity.

KONTRON ELEKTRONIK GMBH

Germany – Freisinger Strasse 21, D-8057 Eching; phone: 08165-77-102; fax: 08165-77-113

66 Cherry Hill Drive, Suite 2000, Beverly, MA01915; phone: (508) 927-6575; fax: (508) 927-6511

The EPP-80 is a stand-alone universal programmer, when used with the Universal Programming Module. The 40-pin socket allows most 20 and 24-pin PLDs to be programmed, and an expanding range of 40-pin devices is also supported.

LOGICAL DEVICES INC.

USA – 1201 NW 65th Place, Fort Lauderdale, FL33309; phone: (305) 974-0967; fax: (305) 974-8531

Europe (software) – CAD Solutions GmbH, Leopoldstrasse 28a/II, 8000 Munchen 40, Germany; phone: 089-89349628

Europe (programmers) – GSH-System Technik, Ebenboeckstrasse 20, 8000 Munchen 60, Germany; phone: 089-8343047; fax: 089-8340448

CUPL is a universal logic compiler which accepts input as Boolean equations, truth table or state diagram, and interfaces to most standard schematic capture packages. Extensive shorthand features, such as variable and bit-field grouping make data entry very efficient. Design is device independent with automatic device selection and partitioning/fitting. All standard PLDs and many LSI types are supported.

ALLPRO is a universal programmer with capability of up to 88 pins. It is driven by a PC expansion slot interface and supports a wide range of standard and LSI PLDs. The basic system is low cost, but upgradable to cover the full device range.

MICROPROSS

France – Parc d'Activité des Pres, 5 Rue Denis-Papin, 59650 Villeneuve d'Ascq; phone: 01-20-47-90-40

UK – CPL, Enterprise House, Station Road, Sawbridgeworth, Herts CM21 9JX; phone: 0279 600313

Germany – Macrotron AG, Stahlgruberring, 8000 Munchen 82; phone: 089 420080

Micropross offer two systems; the ROM3000B is a PC-based programmer supporting a large range of PLDs up to 40 pins. It also supports on-board programming. The ROM5000B comes with its own keyboard, screen and floppy disk drive; this allows on-board program storage and some logic compilation ability. The range of PLD coverage appears to be less than the ROM3000B.

MINC INC

USA – 6755 Earl Drive, Colorado Springs, CO80918; phone: (719) 590-1155; fax: (719) 590-7330

Japan – JMC, Hakusan Hi-Tech Park, 801-1 Hakusan-Cho Midori-Ward, Yokohama City 226

PGA Designer is a universal logic compiler for PLDs as well as FPGAs. Data entry is via logic equations, truth tables, state machine diagrams, waveform specification or schematic capture interface. Simulation is performed at logic and final device level, after optimisation. PLD designs are compiled directly into the final device, FPGA designs are checked for conformity to the selected device and the appropriate format generated for completion by vendor software. PLDesigner is similar system without FPGA support.

OrCAD

USA – 3175 NW Aloclek Drive, Hillsboro, Oregon 97124-7135; phone: (503) 690-9881; fax: (503) 690-9891

UK – ARS Microsystems, Herriard Business Centre, Alton Road, Herriard, Basingstoke, Hants RG25 2PN; phone: 0256 381687

OrCAD/PLD is a logic compiler which supports most 20 and 24-pin PLDs, and is being extended to LSI devices. Data entry is by Boolean equations, state diagram, truth table or schematic entry by OrCAD/SDT3. Logic reduction is performed before compilation to a JEDEC file. OrCAD/PLD will interface to other OrCAD products, such as the OrCAD/VST simulator, in order to make an integrated logic design package.

SMS MICRO SYSTEMS

Germany – Im Morgenthal, D-8994 Hergatz-Schwarzenberg; phone: 07522 4460; fax: 07522 8929

UK – Pronto Electronic Systems Ltd., City Gate House, 399-425 Eastern Avenue, Gants Hill, Ilford, Essex IG2 6LR; phone: 081 554 6222; telex: 8954213; fax: 081 518 3222

USA – Encore Technology Corp., 13720 Midway Road, Suite 105, Dallas TX75244; phone (214) 233-3122; fax: (214) 233-2614

The Sprint Plus and Sprint Expert are PC-based universal device programmers, driven from an expansion card. The Expert has a 40-pin socket and supports most PLDs up to 40 pins; the Plus features a 28-pin socket and a more restricted range of device coverage. A basic logic compiler is supplied with the programmer, accepting logic equations and producing JEDEC files. A useful feature is the ability to upload a design and recompile this as a different PLD, e.g. 16V8 to 18CV8 or vice versa.

STAG PROGRAMMERS LTD.

UK – Martinfield, Welwyn Garden City, Herts AL7 1JT; phone: 0707 332148; fax: 0707 371503

USA – 1600 Wyatt Drive, Santa Clara, CA95054; phone: (408) 988-1118

Japan – Teksel Co. Ltd., Kanagawa Science Park, R&D C-4F, 100-1 Sakado, Takatsu-ku, Kawasaki 213; phone: 044 812-7430

Stag manufacture a range of stand-alone programmers. The System 3000 is a powerful universal programmer with built in keyboard and screen which allows fuse map and, for FPLAs and FPLSs, truth table editing and entry. Device coverage is almost universal up to and including 40 pins, and new architectures and alternative package styles are catered for by a series of expansion modules.

The ZL30B is a dedicated PLD programmer which supports most 20, 24 and 28-pin PLDs. A series of expansion modules allow variations in technology and package style.

Stag also act as UK representatives for the LOG/iC design software.

SYSTEM GENERAL CORP.

Asia/Europe – 3F No. 1, Alley 8, Lane 45, Bao Shing Road, Shin Dian, Taipei, Taiwan, ROC; phone: 886-2-917-3005; fax: 886-2-911-1283

USA – 244 South Hillview Drive, Milpitas, CA95035; phone: (408) 263-6667; fax: (408) 262-9220

UK agent – Kontron Elektronik Ltd., Blackmoor Lane, Croxley Centre, Watford, Herts WD1 8XQ; phone: 0923 245991; fax: 0923 54118

The SGUP-85 is a stand-alone universal programmer which supports PLDs up to 28 pins. The TURPRO-1 is a universal programmer controlled by a PC via the serial port. A standard 40-pin socket adaptor covers devices up to and including 40 pins, while surface mount adaptors up to 128 pins are available for more complex LSI PLDs.

VALID LOGIC SYSTEMS INC.

USA – 2820 Orchard Parkway, San Jose, CA95134; phone: (408) 432 9400; fax: (408) 432 9430

UK – Valid House, 39 Windsor Road, Slough, Berks SL1 4TN; phone: 075 382 0101; fax: 075 370057

Japan – Tokyu Building 2nd Floor, 2-16-8 Minami-Ikebukuro, Toshima-ku, Tokyo 171; phone: 033 980-6421; fax: 033 981-8775

SystemPLD and SystemPGA are work station based design tools for PLDs and FPGAs respectively. Both compilers are integrated into the Valid logic design environment, Logic Workbench. They accept design data as equations, state diagrams, HDLs, waveforms, logic primitives or schematic capture. After optimisation and device selection, the simulator, RapidSIM, will include a worst case timing analysis to give a full picture of final device performance. The schematics can be redrawn in PLD format and the design incorporated into a system PCB layout with the Allegro layout program.

VIEWLOGIC SYSTEMS INC.

USA – 293 Boston Post Road West, Marlboro, MA01752; phone: (594) 480-0881

UK – Daneshill House, Chineham Court, Lutyens Close, Chineham, Hants RG24 0UL; phone: 025 651133

Japan – Marubeni Hytech Co. Ltd., Marubeni Hytech Building, 20-22 Koishikawa 4-Chome, Bunkyo-ku, Tokyo 112; phone: 033 817-4881

ViewPLD is a compiler which is aimed at work stations, although it will run on 386/486 PCs. Data entry is by VHDL, equations, truth tables, state diagrams or schematic capture, or it will accept files from other logic compilers. Compilation proceeds after optimisation and partitioning and the design is then subject to a full timing simulation. Retargeting to an ASIC is possible and the design can be incorporated in the simulation of a full system.

Appendix 3
Further Reading

It would be advisable to read the data book of the appropriate device manufacturer (see Appendix 1) before embarking on a design with one of their products. Many manufacturers offer extensive applications information, even going into details of PCB layout, and advice on the use of design software (see Appendix 2). The following references enlarge on the topics covered in this book:

CHAPTER 1:

1.1 Grove, A.S. (1967) *Physics and Technology of Semiconductor Devices.* 1st ed. New York: John Wiley.
1.2 The staff of Texas Instruments (1989) *TTL Data Book Volume 1.* Published by Texas Instruments.
1.3 The staff of Texas Instruments (1988) *High Speed CMOS Logic Data Book.* Published by Texas Instruments
1.4 Read, John W. (ed) (1985) *Gate Arrays.* 1st ed. London: Collins
1.5 Ghandi, S.K. (1968) *The Theory and Practice of Microelectronics.* 1st ed. New York: John Wiley

CHAPTER 2:

2.1 Any standard logic textbook.

CHAPTER 3:

3.1 Nelmes, Guy (1985) 'The technology of a 1Mbit CMOS EPROM', *New Electronics*, **18**, 22, 70–74
3.2 Mullard Ltd. (1979) *A 16k PROM – Its Design and Application.* Mullard Technical Note 111, TP1705
3.3 Watts, Tony (1992) 'Technology steps to build a 4Mbit EPROM', *Electronic Product Design*, January, 38–42
3.4 E. Hamdy *et al.*, (1988) 'Dielectric based antifuse for logic and memory ICs' *IEDM Tech. Dig.* (San Francisco, CA), 786–789

CHAPTERS 4–5:

Further reading for these chapters is probably confined to the Data books mentioned above

CHAPTER 6:

6.1 Koelling, T.K. (1988), 'Programmable-AND/allocatable-OR based EPLD addresses the needs of complex combinational and sequential designs.' *Wescon/88 Conference Record*, November, p. 22.4/1–8

6.2 Saffari, B. and Tarverdians, F. (1988), 'PML logic design and applications' *Wescon/88 Conference Record*, November, p. 22.2/1–5

6.3 El Gamal, A. *et al.* (1989), 'An architecture for electrically configurable gate arrays.', *IEEE Journal of Solid-State Circuits*, April, **24**, 2, 394–8

6.4 Rajsmun, R. (1989), 'Design of reprogrammable FPLA.', *Electronics Letters*, May, **25**, 11, 715–16

6.5 Bursky, D. (1989), 'Get gate-array pliability with programmable chip.' *Electronic Design*, October, **37**, 22, 127–9

6.6 Jigour, R.J. (1990) 'Optimizing high density PLD design with PEEL arrays and PLACE development software.', *Wescon/90 Conference Record*, November, p. 366–8

CHAPTER 7:

7.1 Schulze, B. (1990), 'Applying design synthesis to programmable logic.', *Wescon/90 Conference Record*, November, p. 375–80

7.2 Minato, K. *et al.* (1991), 'An efficient heuristic approach to state encoding for PLA-based sequential circuits.', *IEICE Transactions*, **E74**, 1, 14–21

7.3 Small, C.H. (1991), 'FPGA design methods', *EDN*, August, **36**, 16, 114–18

CHAPTER 8:

8.1 Holley, M. and Pellerin, D. (1990), 'Design tools keep pace with complex devices.', *Wescon/90 Conference Record*, November, p. 369–74

8.2 Conner, D. (1990), 'Design tools smooth FPGA configuration.', *EDN*, June, **35**, 12, 49–58

8.3 Leibson, S.H. (1990), 'PLD development software.' *EDN*, August, **35**, 16 100–116

8.4 Merrill, H.W. (1990), 'PLD software tools for ASIC prototyping.', *Wescon/90 Conference Record*, November, p. 366–8

APPENDIX 2:

The staff of Advanced Micro Devices Inc., (1991), *Fusion PLD Catalog.* Published by AMD, Sunnyvale CA, USA

Appendix 4
Trade Marks

The following trade marks and registered trade marks were used in this book:

TangoPLD, Tango-Schematic attributed to Accel Technologies
ACT, Action Logic, Activator, PLICE attributed to Actel Corporation
HAL, PAL, PALASM, MACH attributed to Advanced Micro Devices
A+PLUS, MAX+PLUS, MAX attributed to Altera Corporation
Atmel-ABEL attributed to Atmel Corporation
ABEL-4 PLD, ABEL-FPGA, FutureNet; UniSite attributed to Data I/O
  Corporation
EF-PER, ZIFPAC attributed to Elan Digital Systems
AdET 1.0 attributed to Exel Microelectronics
ALL-03 attributed to Hilo System Research Co.
Speedmaster 1000 attributed to ICE Technology Ltd.
iPLSII, 80386, 80486 attributed to Intel Corporation
IBM, PC, OS/2 attributed to International Business Machines Corporation
APEEL, PEEL, PLACE attributed to International CMOS Technology Inc.
LOG/iC attributed to ISDATA GmbH
PROMAC attributed to Japan Macnics Corporation
EPP-80 attributed to Kontron Elektronik GmbH
GAL, Generic Logic Array attributed to Lattice Semiconductor Corporation
CUPL, ALLPRO attributed to Logical Devices Inc.
MSDOS attributed to Microsoft Corporation
PGADesigner, PLDesigner attributed to Minc Inc.
MAPL, OPAL, TRI-STATE attributed to National Semiconductor Corpor-
  ation
OrCAD/PLD, OrCAD/SDT3, OrCAD/VST attributed to OrCAD
AMAZE, PML, SNAP attributed to Philips Semiconductors
PLUS Array, PLUSTRAN attributed to Plus Logic Inc.
System 3000, ZL30B attributed to Stag Programmers Ltd.
TURPRO-1 attributed to System General Corporation
TI-ALS, TPC10, TPC12 attributed to Texas Instruments Inc.
SystemPLD, SystemPGA, Allegro attributed to Valid Logic Systems Inc.
ViewPLD attributed to Viewlogic Systems Inc.
LCA, Logic Cell, XACT attributed to Xilinx Inc.

Index

nibble, 47
NOR gate, 220
npn bipolar transistor, 6

octal counter, 37
  Karnaugh maps (H & L), 42
  Karnaugh maps (toggle), 43
one-time programmable PROMs, 65
open collector output, 223
operating systems, 202
OR function, 2
OR gate, 219
OR matrix, 73
  circuit diagram, 78
output macrocell, 104
output register, 39, 122

p-channel MOS transistor, 5
p-n junction, 4
p-type semiconductor, 3
PA family, 175
package limitations, 184
packages, 8
PAL (programmable array logic), 82
  asynchronous, 108
  basic architecture, 84
  combinational range, 88
  counter design in, 99
  design methods, 85
  emulation by GAL, 104
  enhanced range, 113
  exclusive-OR, 100
  numbering system, 86
  output drive, 107
  registered, 96
  registered range, 103
  zero-power, 107
PAL22IP6, 119
PAL22V10, 105
PAL23S8, 114
PAL32VX10, 117
PALASM, 211
PALCE29M(A)16, 117
parametric tests, 191
parity generator, 27
partitioning, 184
PC based programmer, 213
performance limitations, 182
PGA (programmable gate array), 164
photo-etching, 6
PIA (programmable interconnect array), 150
pipelined operation, 90
PLA (programmable logic array), 73
  range, 82
place and route, 209

planar process, 6
plastic leaded chip carrier, 9
PLC42VA12, 134
PLD selection guide, 183
PLE (programmable logic element), 68
  bipolar range, 68
  CMOS range, 69
PLICE (programmable low impedance circuit element), 57
PLS (programmable logic sequencer), 120
  range, 127
PLS105, 134
PLS155, 126
Plus Logic, 152
PML (programmable macro logic), 159
pnp bipolar transistor, 6
power up state, 125
preload, 199
present state, 40
  software definition, 206
product term, 76
  sharing, 117
  variable, 105
production programming, 213
  external vs. in-house, 214
programmability, 62, 188
  batch effects, 188
  statistical fluctuations, 189
programmable flip-flop type, 110
programmable output polarity, 73
programming algorithm, 212
programming specification, 189
programming yield, 188
  statistical fluctuations, 189
PROM (programmable read only memory), 50
  as logic element, 66
  correspondance with Karnaugh map, 67
  diagnostic, 94, 160
  programming circuit, 52
  registered range, 95
  serial, 166
protocols, 262
prototyping with PLDs, 217
pseudo-random number generator, 250

Quine–McCluskey method, 207

R-S latch, 232
race conditions, 197
RAM (random access memory), 165, 266
register, 35
  preload, 200